ALGEBRA
FOR
EVERYONE

edited by

EDGAR L. EDWARDS, JR.

Virginia Department of Education

Richmond, Virginia

MATHEMATICS EDUCATION TRUST

established by the

NATIONAL COUNCIL OF TEACHERS OF MATHEMATICS

Third printing 1993

Library of Congress Cataloging-in-Publication Data:

Algebra for everyone / edited by Edgar L. Edwards, Jr.
p. cm.
Includes bibliographical references.
ISBN 0-87353-297-X
1. Algebra—Study and teaching. I. Edwards, Edgar L.
II. Mathematics Education Trust.
QA159.A44 1990 90-6272
512'.071—dc20 CIP

The publications of the National Council of Teachers of Mathematics present a variety of viewpoints. The views expressed or implied in this publication, unless otherwise noted, should not be interpreted as official positions of the Council.

Financial support for the development of this publication has been provided by the Julius H. Hlavaty and Isabelle R. Rucker endowments in the Mathematics Education Trust (MET). The MET is a foundation established in 1976 by the National Council of Teachers of Mathematics. It provides funds for special projects that enhance the teaching and learning of mathematics.

Printed in the United States of America

TABLE OF CONTENTS

Acknowledgments iv

Preface v

1. The Problem, The Issues That Speak to Change 1
   DENE R. LAWSON, California State Department of Education (retired), Sacramento, California

2. Prior Experiences 7
   HILDE HOWDEN, Albuquerque, New Mexico

3. The Transition from Arithmetic to Algebra 24
   RICHARD LODHOLZ, Parkway School District, Saint Louis County, Missouri

4. Enhancing the Maintenance of Skills 34
   DAVID J. GLATZER, West Orange Public Schools, West Orange, New Jersey

   GLENDA LAPPAN, Michigan State University, East Lansing, Michigan

5. Teacher Expectations of Students Enrolled in an Algebra Course 45
   ROSS TAYLOR, Minneapolis Public Schools, Minneapolis, Minnesota

6. Instructional Strategies and Delivery Systems 53
   FRANKLIN D. DEMANA, Ohio State University, Columbus, Ohio

   BERT K. WAITS, Ohio State University, Columbus, Ohio

7. Communicating the Importance of Algebra to Students 62
   PAUL CHRISTMAS, John Hersey High School, Arlington Heights, Illinois

   JAMES T. FEY, University of Maryland, College Park, Maryland

8. List of Resources 74
   DOROTHY S. STRONG, Chicago Public Schools, Chicago, Illinois

Acknowledgments

THE Mathematics Education Trust Committee (MET) wishes to acknowledge the many individuals who contributed to this publication either as an author, consultant, or typist or in an editorial capacity. Special appreciation goes to the authors, whose names appear in the Table of Contents. Their fine essays are a tribute to their understanding of the national need to provide a strong comprehensive mathematics curriculum for all students.

The individuals who met in Chicago at the NCTM 1988 Annual Meeting offered much guidance and assistance to the MET Committee as it formulated its ideas regarding the format and approach to the development of this document. These persons, who met with MET Committee members Stuart A. Choate, Edgar L. Edwards, James D. Gates, Patricia M. Hess, and Margaret J. Kenney, are—

David Glatzer, West Patterson, N.J.
Harriet Haynes, Brooklyn, N.Y.
Hilde Howden, Albuquerque, N.Mex.
Marie Kaigler, New Orleans, La.
Richard Lodholz, Saint Louis, Mo.
Dorothy Strong, Chicago, Ill.
Ross Taylor, Minneapolis, Minn.
Judith Trowell, Little Rock, Ark.
David Williams, Philadelphia, Pa.
Leslie Winters, Northridge, Calif.

The Mathematics Education Trust Committee is most appreciative of the work of Edgar L. Edwards, who posed the problem, formulated the original question, kept the discussion alive, and shepherded the entire project from beginning to end. He compiled the essays, and the typing of consistent copy was done by his secretary, Joy Hayes. Her work was a great deal of help to those who read and edited this document for publication.

The Mathematics Education Trust Committee expresses appreciation for the work and effort of Albert P. Shulte, who served as an external editor. He reviewed the document, edited it for consistency, and put forth much time and effort in its preparation. Finally, but not least, is the work of Stuart A. Choate, chair of the Mathematics Education Trust Committee at the time the publication was completed. He also reviewed the entire document and made a significant contribution to the publication.

Through the efforts of many people this publication is able to address the critical need for mathematics educators to provide algebra for everyone in this technological society as we continue our present pathway in business and industry.

The Mathematics Education Trust Committee

Preface

THE Mathematics Education Trust (MET) Committee began a discussion of critical areas of need in mathematics education, and algebra became a focal point of deliberation. It was recognized that algebra, considered as a course, is important to persons desiring a career in such specialized areas as engineering. However, one must take a much broader view after considering the many national reports regarding education, especially the NCTM's *Curriculum and Evaluation Standards for School Mathematics* and the publication *Everybody Counts* (Washington, D.C.: Mathematical Sciences Education Board and National Research Council, 1989). The two named documents clearly indicate that algebra must be included in the teaching of mathematics on a much broadened scale in addition to the formal course currently called "algebra."

The fundamentals of algebra and algebraic thinking must be part of the background of all our citizens who are in the workplace, all who read the news, and those who wish to be intelligent consumers. The vast increase in the use of technology in recent years requires that school mathematics ensure the teaching of algebraic thinking as well as its use at both the elementary and the secondary school levels. This new technology presents opportunities to generate many numerical examples, to graph data points, and to analyze patterns and make generalizations regarding the information at hand.

Business and industry are requiring of their employees higher levels of thinking that go beyond those acquired in a formal course in algebra. Of great concern to the membership of NCTM, as well as business and industry, is the future of our low-achieving and underserved students. These segments of our population must be given the necessary algebraic background, beginning at the elementary school level, so that they can either engage in the formal course or be able to compete in the marketplace, where general algebraic concepts and skills are necessary.

Algebra for Everyone is a set of essays, each pertaining to a specific aspect of the need to teach the fundamentals of algebra to the entire population. All students need to develop systematic approaches to analyzing data and solving problems. Algebra is a universal theme that runs through all of mathematics, and it is a tool required by nearly all aspects of our nation's economy. For many young people, algebra is perceived to be an entirely separate branch of mathematics with no relation to what was learned in earlier grades. In the early grades, students must be helped to make connections among mathematical ideas and to build relationships between arithmetic and algebra.

At the NCTM's 1988 Annual Meeting in Chicago, the MET Committee invited a group of supervisors to discuss this topic and to suggest ways in which this need for a broader understanding of algebra by our entire populace might be attained. From that meeting, *Algebra for Everyone* began. This document is written for supervisors and teachers who influence the teaching of algebra and who can and will address the problems facing our underserved and underachieving populations. The conclusion of the MET Committee was that the principles educators should follow for the populations just mentioned are also valuable principles for our total population in general, including those who will be enrolled in a formal algebra course. If these principles are addressed across the broad curriculum, from the early grades on, understandings will increase, and the underserved and underachieving will achieve and be served. Thus this booklet does not focus specifically on these groups; rather, it is for all students. The fundamentals of algebra are necessary for all students if they are going to be successful in the job market and if the business and industrial community is to maintain its high level of success around the world.

Mathematics Education Trust Committee, 1987–90

Jesse A. Rudnick
Edgar L. Edwards
Patricia M. Hess
Stuart A. Choate
Margaret J. Kenney
Harold D. Taylor
Bruce C. Burt
James D. Gates

1

THE PROBLEM, THE ISSUES THAT SPEAK TO CHANGE

Dene R. Lawson

ALGEBRA FOR EVERYONE?

The time has come to explore the possibilities for a new generation of students who deserve a better mathematics curriculum. The study of algebra is a key element in understanding mathematical systems and should not await high school freshmen—or precocious eighth graders—as if they are required to master computation before being introduced to algebraic concepts. For example, learning algebraic concepts through concrete models is well within the intellectual grasp of primary-aged students.

STAYING CURRENT IN A TECHNOLOGICAL SOCIETY

During the past thirty years, an explosion of knowledge has created obsolescence in our educational system. Advancements in space-age technology have quadrupled the mathematical and scientific knowledge that we knew in 1950. Only one human lifetime has elapsed since the Wright brothers flew the first powered and controlled airplane in 1903. We marvel that Neil Armstrong and eleven other astronauts have walked on the moon. Even the skeptics are aware that the hand-held calculator and the computer are here to stay. Thanks to the relatively inexpensive microchip and other space-age inventions, our society is moving at a faster pace in every way.

Unfortunately, many of our students have difficulty functioning in a technological society. Our present educational system is falling farther behind. A basic education must go beyond reading, writing, and arithmetic to encompass communication, problem-solving skills, and scientific and technological *literacy*. We need to train our students to enter the twenty-first century with the capacity to understand the technological society.

In simple applications of computing skills or in problem-solving situations, United States students are well below the international average. In fact, with the highest average classroom size of forty students, Japan obtained the highest average achievement scores of all twenty countries that participated in the Second International Mathematics Study, as reported in McKnight et al. (1987). At the same time, with an average class size of twenty-six students, the average scores of United States eighth-grade classrooms among the twenty countries in different subject areas were tenth in arithmetic, twelfth in algebra, sixteenth in geometry, and eighteenth in measurement.

It is worth noting that in Japan algebra tends to drive the mathematics curriculum. By contrast, most students in the United States have spent an inordinate amount of their schooling trying to learn the skills of paper-and-pencil arithmetic. Even at age seventeen, these students do not possess the breadth and depth of mathematics proficiency needed for advanced study in secondary school mathematics.

Without mathematics "know-how," many students will qualify for only marginal employment. Mathematics education is more than learning to compute. As Pólya (1962, pp. vii, viii) advised,

> Our knowledge about any subject consists of *information* and *know-how*. If you have genuine *bona fide* experience of mathematical work on any level, elementary or advanced, there will be no doubt in your mind that, in mathematics, know-how is much more important than mere possession of information.... What is know-how in mathematics? The ability to solve problems—not merely routine problems but problems requiring some degree of independence, judgment, originality, creativity.

Opportunities to learn are not committed or distributed fairly among United States students. By the seventh grade, classes and topics have become quite differentiated, as students who are perceived to be weak in mathematics are moved to remedial groups or given less challenging subject matter. Some students are forever relegated to a slower pace or to different courses, depriving them of learning opportunities accorded other students. As Dossey et al. (1988, p. 9) report,

> Too many students leave high school without the mathematical understanding that will allow them to participate fully as workers and citizens in contemporary society. As these young people enter universities and business, American college faculty and employers must anticipate additional burdens. As long as the supply of adequately prepared precollegiate students remains substandard, it will be difficult for these institutions to assume the dual responsibility of remedial and specialized training; and without highly trained personnel, the United States risks forfeiting its competitive edge in world and domestic markets.

A technological society increasingly needs professional users of mathematics. The Task Force on Women, Minorities, and the Handicapped in Science and Technology (1988) reports to the president, the Congress, and the American people that our educational pipeline—from kindergarten through the doctoral level—is failing to produce the workers needed to meet future demand in the scientific and engineering work force.

Sadly, some teachers and counselors tend to categorize students' ability by nonintellectual criteria, such as color of skin, wearing apparel, physical disability, or sex. This discrimination leads to differential treatment in the classroom or in counseling, which often creates real differences in students' performance. Students who believe that they cannot learn also participate in selecting less difficult subjects to study.

In the year 2000, 85 percent of new entrants to the work force will be members of minority groups and women. The number of people with disabilities who can function adequately is also on the rise. These underrepresented groups must be encouraged to seek positions in science and engineering, because without additional human resources, a critical future shortfall of scientists and engineers is certain (Task Force on Women, Minorities, and the Handicapped in Science and Technology 1988, p. 3). A pluralistic, democratic society recognizes that *all* its citizens must have an equitable opportunity to share in its benefits.

Arithmetic skills are important but not to the extent that we have previously demanded. Basic arithmetic facts are still necessary elements of the well-educated mind, but being able to estimate mentally the approximate answer to 78 times 62 is just as important as being able to find the answer with paper and pencil. The calculator works faster and more accurately, but we still need the kinds of skills that supplement the frequent use of the calculator. We still need confidence in our results.

THE NEW VISION: RECOMMENDED CHANGES FOR MATHEMATICS EDUCATION

Social circumstances change, especially in a technological society, and its schools must try to stay current. Thus, another revolution in mathematics education is going on today, inspired by such documents as *A Nation at Risk* (National Commission on Excellence in Education 1983) and *An Agenda for Action* (National Council of Teachers of Mathematics 1980). The basic principles for mathematics teachers espoused by the late George Pólya in the 1960s closely match the philosophy of this new vision. In a series of institutes funded by the National Science Foundation, Pólya, revered Stanford mathematics professor, taught hundreds of high school teachers his philosophy (1962, p. viii):

a) The first and foremost duty of the high school in teaching mathematics is to emphasize *methodical work in problem solving.*

b) The teacher should know what he is supposed to teach.

c) The teacher should develop his students' know-how, their ability to reason.

How Do We Turn Things Around?

Several recommendations for restructuring the mathematics curriculum are reported in *The Underachieving Curriculum* (McKnight et al. 1987, pp. xii–xiii); they are based on a national report on the Second International Mathematics Study. One recommendation underlines the importance of providing algebra for everyone—the theme of this booklet:

> Concerning substance, the continued dominating role of arithmetic in the junior high school curriculum results in students entering high school with very limited mathematical backgrounds. The curriculum for all students should be broadened and enriched by the inclusion of appropriate topics in geometry, probability and statistics, as well as algebra.

No student should be denied an opportunity to learn the skills that a technological society demands for survival. Clearly, hand-held calculators and desk-top computers are now available in sufficient numbers for classroom instruction. Schools are increasingly committing resources to the hardware and supplementary materials that take advantage of this relatively new technology.

Probably the greatest value in using calculators as a classroom tool is the vast amount of time that is liberated when we no longer assign hundreds of practice problems that typify the current mathematics curriculum. Calculators are a major breakthrough in teaching more complex concepts (e.g., exponential functions, series, sequences, iterations) by eliminating peripheral arithmetic that in the past used up most of the available time.

In the adult world, calculators have caught on quickly. If a calculator is handy, few adults use paper and pencil for balancing a checkbook. For adults, knowing *when* to divide has become a more important skill than knowing *how*. It is no longer necessary to develop paper-and-pencil proficiency with large numbers. With the calculator as a learning tool, students can use the newly found time to develop their abilities to use information creatively: guessing, iterating, formulating, and solving.

The same message applies to the use of the computer. Its expense in relation to the calculator may limit its availability, but we assume that computers and appropriate software will become increasingly accessible for classroom use. The paper-and-pencil practice of graphing a nonlinear function point by point to examine certain properties of the function, for example, will not be necessary once the definition of a function and its coordinate representation are understood. Computer technology makes it possible to explore concepts associated with functions, such as domain, range, maxima and minima, and asymptotes. Students will be able to observe the behavior of a variety of functions, including linear, quadratic, general polynomial, and exponential, without succumbing to the drudgery of calculations that in the past have been paper-and-pencil tasks.

A commission representing the National Council of Teachers of Mathematics has prepared *Curriculum and Evaluation Standards for School Mathematics* (NCTM 1989). This document is a comprehensive effort to provide a standard source for mathematics educators to use in making changes that reflect the new vision.

Each curriculum standard at three different grade spans describes—

a) the mathematical content to be learned;

b) expected student actions associated with that content;

c) the purpose, emphasis, and spirit of this vision for instruction.

Eleven curriculum standards for grades 5–8 have been identified and elaborated. They are mathematics as problem solving, mathematics as communication, mathematics as reasoning, number, number systems, computation and estimation, measurement, geometry, statistics, probability, and algebra.

Several of the standards are relevant to this booklet, but the following example will serve to illustrate this link (NCTM 1989, p. 102):

STANDARD 9: ALGEBRA

In grades 5–8, the mathematics curriculum should include explorations of algebraic concepts and processes so that students can—

- understand the concepts of variable, expression, and equation;
- represent situations and number patterns with tables, graphs, verbal rules, and equations and explore the interrelationships of these representations;
- analyze tables and graphs to identify properties and relationships;
- develop confidence in solving linear equations using concrete, informal, and formal methods;
- investigate inequalities and nonlinear equations informally;
- apply algebraic methods to solve a variety of real-world and mathematical problems.

For illustrative purposes, only the fifth of the fourteen curriculum standards for grades 9–12 is printed here (NCTM 1989, p. 150), although several standards relate to algebra:

STANDARD 5: ALGEBRA

In grades 9–12, the mathematics curriculum should include the continued study of algebraic concepts and methods so that all students can—

- represent situations that involve variable quantities with expressions, equations, inequalities, and matrices;
- use tables and graphs as tools to interpret expressions, equations, and inequalities;
- operate on expressions and matrices, and solve equations and inequalities;
- appreciate the power of mathematical abstraction and symbolism;

and so that, in addition, college-intending students can—

- use matrices to solve linear systems;
- demonstrate technical facility with algebraic transformations, including techniques based on the theory of equations.

Through the influence of recent national and state documents that recommend the "new vision" in mathematics education, significant changes are occurring in textbooks, standardized tests, supplementary materials, access to calculators and computers, and in state and local commitment to staff development.

Staff Development Is the Key to Success

Obviously, a strong commitment to staff development is essential because without it, teachers tend to teach the way they have been taught. To make a difference, we need teachers who are ready to make changes. However, teaching for *understanding* frequently means using concrete models to introduce a new concept rather than assuming that students understand the abstract or strictly symbolic model of a concept. For teachers who insist on a steady pace, momentary delays for understanding are troublesome. Teachers will need help in learning how to budget their time in different ways, and in many instances, they will have to practice with concrete models and vary their presentation to enhance students' understanding.

School administrators must supply updated textbooks, correlated supplementary materials, and the resources for in-service opportunities that demonstrate the teaching of know-how. With support from their principals, teachers will be confident that their own renewed commitment is appreciated. Although a principal may not be strong enough in mathematics to train or advise the teachers, his or her commitment to program success will inspire teachers to team up or develop their own strategies to improve mathematics instruction. Without official encouragement, it may be "business as usual," an outcome that the target students can ill afford.

ALGEBRA FOR EVERYONE

This booklet is designed to support the teaching of algebra to *all* students. In a simplistic way, algebra may be described as generalized arithmetic. For elementary algebra, that definition is probably sufficient. At some point, eighth or ninth grade, a more formal approach to the study of algebra begins. The earlier exposure to some aspects of algebra in elementary and middle schools should furnish essential background for students.

At an early age, topics will be introduced from all mathematics strands: number, measurement, geometry, patterns and functions, statistics and probability, logic, *and* algebra. By the time students reach the eighth or ninth grade, they will already have confronted many ideas that heretofore have been avoided for totally unfounded reasons. For many students, especially the eventual dropouts, few topics beyond computation are ever introduced, and the world of mathematics seems dull, redundant, and of little use in their future world.

We have learned that our present curriculum leads to mediocrity. We know we can do better. *Every* student deserves a teacher who has made a conscious decision to teach for understanding, to train students to approach problem solving with confidence, and to help students develop number sense, or know-how, along with the mathematical concepts typically expected of them. Every student deserves the opportunity to learn algebra—it is a key element of know-how. In many instances, knowledge of algebra may be the key that unlocks curiosity, creativity, and ambition in the classroom, and later, success in a mathematically oriented technological world.

REFERENCES

Dossey, John A., Ina V. S. Mullis, Mary M. Lindquist, and Donald L. Chambers. *The Mathematics Report Card: Are We Measuring Up? Trends and Achievement Based on the 1986 National Assessment*. Princeton, N.J.: Educational Testing Service, 1988.

McKnight, Curtis C., F. Joe Crosswhite, John A. Dossey, Edward Kifer, Jane O. Swafford, Kenneth J. Travers, and Thomas J. Cooney. *The Underachieving Curriculum: Assessing U.S. School Mathematics from an International Perspective*. Champaign, Ill.: Stipes Publishing Co., 1987.

National Council of Teachers of Mathematics. *An Agenda for Action: Recommendations for School Mathematics of the 1980s*. Reston, Va.: The Council, 1980.

National Council of Teachers of Mathematics, Commission on Standards for School Mathematics. *Curriculum and Evaluation Standards for School Mathematics*. Reston, Va.: The Council, 1989.

National Commission on Excellence in Education. *A Nation at Risk: The Imperative for Educational Reform*. Washington, D.C.: U.S. Government Printing Office, 1983.

Pólya, George. *Mathematical Discovery: On Understanding, Learning, and Teaching Problem Solving*. Vol. 1. New York: John Wiley & Sons, 1962.

Task Force on Women, Minorities, and the Handicapped in Science and Technology. *Changing America: The New Face of Science and Engineering*. Interim Report. Washington, D.C.: The Task Force, 1988.

2

PRIOR EXPERIENCES

HILDE HOWDEN

WITH THE REALIZATION that algebra is for everyone comes the increased need for all students to have the prior experiences necessary for success in the formal study of algebra. Identifying these experiences and suggesting when and how they should best be offered are the concerns of this chapter.

No magic point marks the beginning of preparation for the study of algebra. Preparation for algebra begins with the recognition that numbers can represent a wide variety of quantities, that numbers can be classified and related according to their characteristics, and that these relationships can be communicated in a variety of ways. For example, consider the myriad of algebraic concepts involved when a student "discovers" that for any whole-number replacement of $\square$ (or n) by $2\square$ (or $2n$), an even number always results.

Classification: All whole numbers are either even or odd.

Reasoning: Since whole numbers appear to be alternately odd and even, they form a pattern: O, E, O, E, O, E,....Some characteristic of numbers must account for this pattern.

Number relationships: An even number consists of pairs; an odd number consists of pairs and one extra.

Even Numbers	Odd Numbers
	1 x
2 x x	3 x x x
4 x x x x	5 x x x x x
6 x x x x x x	7 x x x x x x x

Operation sense: When an even number is divided by 2, no remainder results; when an odd number is divided by 2, a remainder of 1 always results.

$\frac{24}{2} = 12$ $\frac{25}{2} = 12$ R: 1

$\frac{346}{2} = 173$ $\frac{347}{2} = 173$ R: 1

Generalization: Every even number is a multiple of 2.

Notion of variable: Every even number is equal to 2 times some number.

Organization of information:

Whole number	1	2	3	4	5	6	7	...
Corresponding even number	2	4	6	8	10	12	14	...

Dynamics of change: As the whole numbers increase by 1, the corresponding even numbers also increase, but they increase by 2.

Concept of implication: If a whole number is even, then it can be expressed as $2n$, where n is a whole number.

Concept of function: The set of even numbers can be generated by multiplying members of the set of whole numbers by 2.

Use of notation: Even numbers can be represented as $2n$, where n represents any whole number. That is, $f(n) = 2n$.

Nature of answer: A mathematical solution is not necessarily expressed as a number. In this example, the solution of how to represent an even number is $2n$.

Justification: The solution can be justified either by substituting whole numbers for n to check that $2n$ is always an even number or by showing that since $2n$ is a multiple of 2, it is an even number according to the foregoing generalization statement.

The foundation for these and similar concepts, traditionally considered to be components of a formal study of algebra, is gradually developed throughout the K–8 mathematics curriculum, as recommended by the NCTM's *Curriculum and Evaluation Standards for School Mathematics* (1989) (hereafter called *Standards*). Thus, an analysis of prior experiences recommended for the study of algebra must encompass the entire spectrum of a student's mathematical experiences prior to the study of algebra.

To identify the most relevant experiences that mold the foundation for algebra, numerous research studies concerning difficulties students encounter in the study of algebra and a variety of algebra prognosis tests and middle school contest examinations were analyzed. Their common expectations are included in this brief summary.

CONTENT AND PROCESS

Although knowledge of specific content and vocabulary is a necessary ingredient of the foundation for algebra, of at least equal importance is the ability to look beyond the numerical details or dimensions to the essence of a situation. This ability is a learned skill; it is not an inherited talent. Developing this ability requires that instruction focus on process as well as content.

The four process standards, common to each of the grade-level designations (K–4, 5–8, and 9–12) in the NCTM's *Standards* (1989), are problem solving, reasoning, communication, and connections. They permeate instruction of all content and will thus be referenced throughout the discussion of the content areas considered in this summary.

Because mathematics is not a compendium of discrete bits and pieces that can be taught independently, its instruction cannot be classified into neatly defined categories. However, five major content categories were selected for consideration in this summary for the best alignment with current research and a majority of state and district curriculum guides for mathematics. They are patterns, relationships, and functions; number and numeration; computation; language and symbolism; and tables and graphs.

The following brief descriptions of the content and process categories illustrate their interdependence, which is further illustrated in a discussion of learning experiences that traces the study of multiples throughout the K–8 curriculum.

Patterns, Relationships, and Functions

Mathematics is often described as the study of patterns. Students who, from the earliest grades, are encouraged to look for patterns in events, shapes, designs, and sets of numbers develop a readiness for a generalized view of mathematics and the later study of algebra. Recognizing, extending, and creating patterns all focus on comparative thinking and relational understanding. These abilities are integral components of mathematical reasoning and problem solving and of the study of specific concepts, such as percentage, quantitative properties of geometric figures, sequence and limit, and function.

By analyzing and creating tables and graphs of data they have recorded, students develop an understanding of the dynamics of change. By modeling increasing and decreasing relationships, students recognize how change in one quantity effects change in another, which is the essence of proportional reasoning. Consider the following activities.

Students work in groups of four, with a designated duty for each student in the group. One student should record the data, another should be a reporter, and two should be explorers. Each group has several geoboards. Half the groups are given forty pieces of construction-paper squares that each measure 1 geoboard unit on a side and a loop of string whose length is exactly twenty-four of the geoboard units. These groups are to determine how many different rectangles they can make by placing the loop of string around nails of the geoboard and then recording the number of construction-paper squares needed to cover the interior of each rectangle.

The other groups are given a large rubber band and thirty-six construction-paper squares. Their job is to use the paper squares to identify all the possible rectangles whose area is thirty-six square units, and then to record the number of units the rubber band spans to enclose each area.

The results differ with the grade level of the students and their prior experience with cooperative learning. In general, however, the recorded data and final reports resemble the following:

Perimeter: 24 Units

l	w	Area
6	6	36
7	5	35
8	4	32
9	3	27
10	2	20

Area: 36 Square Units

l	w	Perimeter
6	6	24
9	4	26
12	3	30
18	2	40
36	1	74

Students in the first group experience how change in the dimensions of the rectangle affects its area. They recognize that the change is not constant but that it follows a pattern. For consecutive unit increases in the length (and consequent unit decreases in the width), the area decreases by consecutive odd numbers.

Students in the second group also experience the dynamics of change. However, they recognize some dramatic differences. As the length increases, the perimeter also increases. The increases are all even numbers, but a pattern does not appear to be evident, at least not an easily recognizable pattern. The students wonder whether using unit increments of change for the length would clarify the pattern.

l	6	7	8	9	10	11	12	13	14
w	6	5.1	4.5	4	3.6	3.3	3	2.8	2.6
P	24	24.3	25	26	27.2	28.5	30	31.5	33.1

The consecutive increases in the perimeter are smaller, but the pattern is still not clear. Perhaps a comparison of the two graphs would help. See figure 2.1. With help, some students will be capable of pursuing the investigation. However, all students will be introduced to algebraic concepts in a problem-solving setting.

Number and Numeration

The NCTM's *Standards* calls it "number sense"; in *Mathematics Counts (The Cockcroft Report)* (1982), the United Kingdom Committee of Inquiry into the Teaching of Mathematics in the Schools calls it "the sense of number"; Bob Wirtz (1974) has referred to it as "friendliness with numbers." Whatever it is called, research has found that this intuition about numbers and how they are related is an important ingredient of learning and later applying mathematics. The NCTM's *Standards* identifies five characteristics of students with good number sense: they have a broad understanding of (1) number meanings, (2) multiple relationships among numbers, (3) relative magnitudes of numbers, (4) the relative effect of operating on numbers, and (5) referents for measures of common objects and situations in their environment.

The development of these characteristics should be an ongoing focus throughout the curriculum to include experiences with whole numbers, fractions, decimals, integers, and irrational numbers. In the early grades, experiences with manipulatives illustrate equivalent forms of numbers. See figure 2.2. In the intermediate grades, explorations with calculators extend this understanding:

$$64 = 8^2,\ 4^3, \text{ or } 2^6$$

$$\sqrt{75} \text{ is between 8 and 9 because } \sqrt{75} \text{ is between } \sqrt{64} \text{ and } \sqrt{81}$$

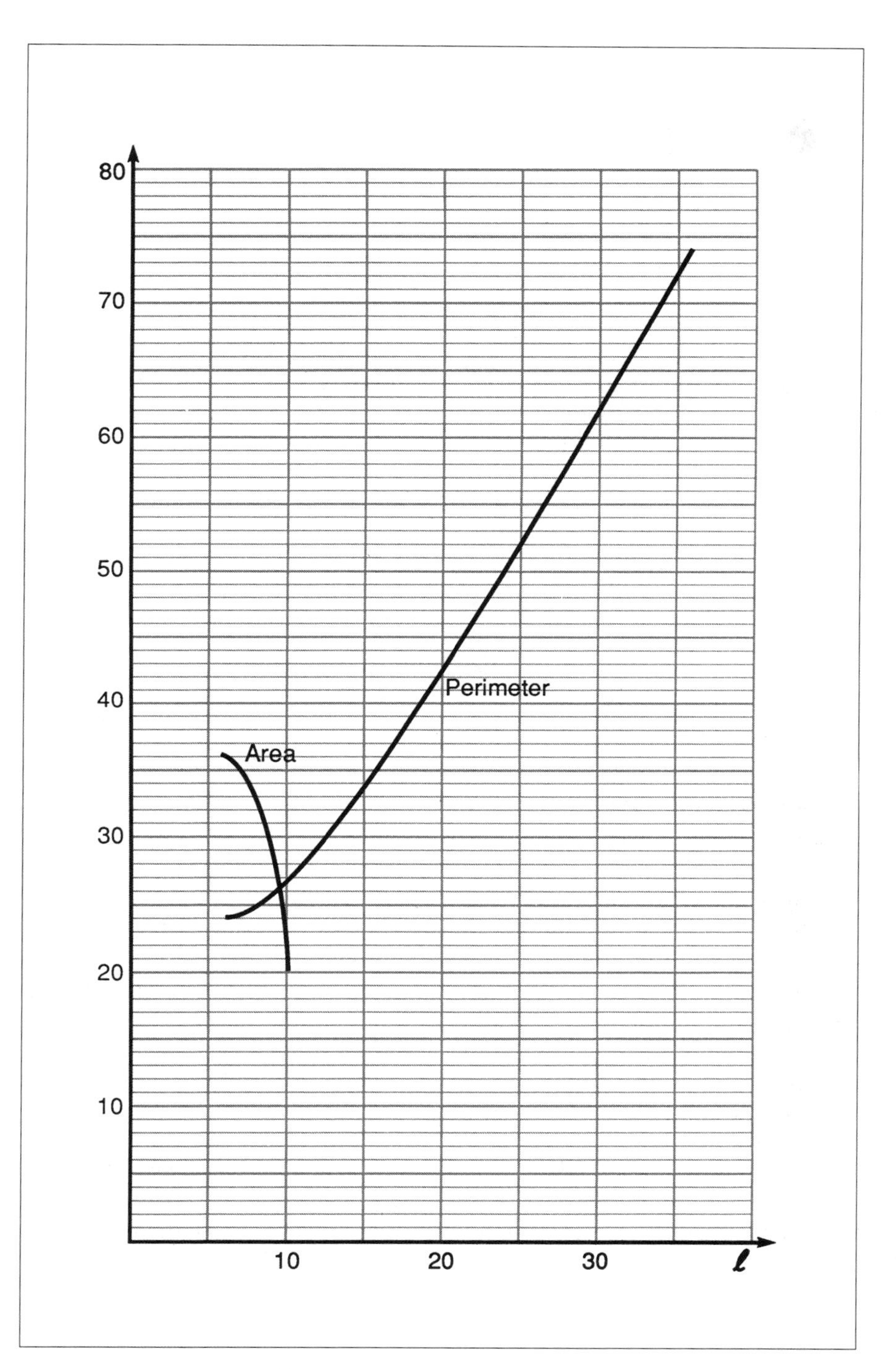

Fig. 2.1. Comparison of collected data

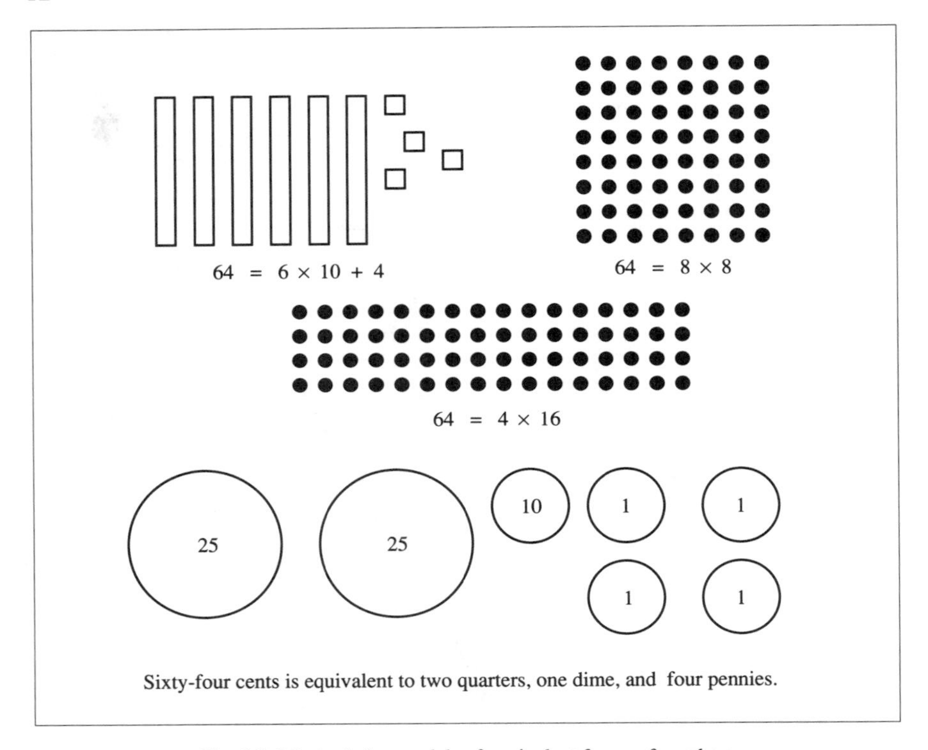

Fig. 2.2. Manipulative models of equivalent forms of numbers

In "Incredible Equations," an activity used in the *Box It and Bag It* program (Burk, Snider, and Symonds 1988), students spend a few minutes each day expressing the day of the month in as many ways as they can devise. The representations become more complex at each grade level as students incorporate into their work new knowledge and experience with numerical operations and symbolism.

As each new number system is studied, students should be given opportunities to acquire a "feel," or "sense," of the numbers in the system; the symbols used to represent them; and their role in the real world; and to discover how operations on them compare with, and differ from, previously studied sets of numbers.

Computation

As reported in *The Mathematics Report Card* (Dossey et al. 1988), the report of the 1986 National Assessment of Educational Progress, nearly one-half of the students at grades 7 and 11 agreed that mathematics is mostly memorizing. More than 80 percent of students at these grade levels viewed mathematics as a rule-bound subject. The fact that other recent research substantiates these findings explains, at least in part, another finding. In a comparison with earlier assessments, the level of students' performance on questions that require application of concepts and problem solving has decreased despite an increase in performance that requires only basic operations with whole numbers. These findings indicate the need for computational skills to be developed in a problem-solving context.

Instructional experiences must require students to model, explain, and develop reasonable proficiency in adding, subtracting, multiplying, and dividing whole numbers, fractions, decimals, and integers. Manipulative models, used first to explore operations with whole numbers, can be extended to fractions, decimals, and binomial expressions, as illustrated in figure 2.3.

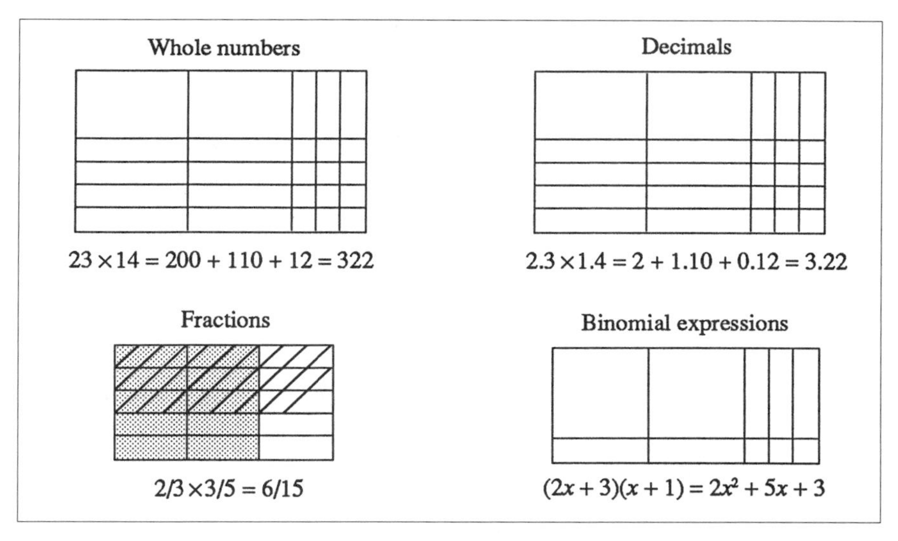

Fig. 2.3. Area model of multiplication

Students should be expected to select and use appropriate methods for computing; they should recognize the conditions under which estimation, mental computation, paper-and-pencil algorithms, or calculator use is most appropriate; and they should be able to check the reasonableness of their results. Such qualitative reasoning—that is, knowing whether or not an answer makes sense—is nurtured by a problem-solving context in which the computation is meaningful to students.

Language and Symbolism

The difficulties that some students encounter with the symbolism of algebra can usually be traced to early misunderstanding of vocabulary and operational symbols used in previous grades. For example, many of the terms used in mathematics have a technical meaning that is very different from the common meaning with which students are familiar. *Volume, foot, plus, value, divide,* and *negative* are just a few examples of words whose technical meanings are frequently not understood by students. Often the stumbling block is the vocabulary, not the mathematics.

Experiences should help students develop multiple meanings for symbols. When the symbol "+" is interpreted to mean only "plus," students who have no difficulty solving an equation of the form 5 + ? = 9 by counting on from 5 to 9, are not sure where to begin counting to solve ? + 5 = 9. Early experiences that introduce positive and negative numbers to represent temperatures above and below zero, altitudes above and below sea level, money earned and spent, yardage gained and lost in a football game, and numbers on either side of zero on a number line help students later to understand the concept of integers.

Even in the middle grades, students benefit from using a balance scale to understand the meaning of "=." Students who have always read "=" as "makes" rather than as "is equal to" have difficulty understanding equations that include variables. For them, a variable has meaning only when its value is known.

To overcome this misconception and to build confidence in using variables, many opportunities should be offered for students to work with variables and equations, two topics that have been identified as presenting great difficulty in the study of algebra. The activity "guess my rule" presents one such opportunity. Given a table of values for two variables, students express the relationship first verbally and then as an open sentence:

p	0	1	2	3	4	5
q	3	5	7	9	11	13

$$q = 2p + 3$$

Alternatively, given a rule like "I am one more than the square of a number," students write the open sentence and then generate a table of values:

x	0	1	2	3	4	5
y	1	2	5	10	17	26

$$y = x^2 + 1$$

Tables and Graphs

As evidenced by the success of *USA Today*, both the newspaper and the television show, we have grown accustomed to communicating information through graphs and tables. Computer capabilities have made such graphic portrayal of information easily accessible. But how many of our students truly understand how to read and interpret information given in this format?

Most of the recently published mathematics textbooks furnish many graphing opportunities at the early grades, but not many carry this emphasis into the middle grades. Yet at this age students enjoy collecting all kinds of things—coins, stamps, posters, rocks, T-shirts, and so on. Experiences in tabulating and graphing such collections should be extended to include graphs of relationships in the coordinate plane, frequency diagrams, scatter graphs, graphs of sample spaces to determine probabilities, spreadsheets, and data base programs. These representations should be used to make inferences and convincing arguments based on data analysis. All such experiences help to relate the dynamic nature of function to everyday occurrences and provide a foundation for visualizing the characteristics of equations and systems of equations to be studied formally in algebra.

Process Skills

Examples of the four process skills—reasoning, problem solving, communication, and connections—permeate the preceding discussion of mathematical content to illustrate both their importance in, and inseparability from, the study of mathematics. Because students do not always recognize when these skills are being used, it is important to focus on their use and to identify them by name. The following brief descriptions summarize the kinds of experiences that are vital to development.

- *Reasoning* is used in making logical conclusions, explaining thinking, and justifying answers and solution processes.
- *Problem solving* is much more than solving word problems. It should be the context within which content is investigated and understood, strategies are developed and applied, and results are interpreted and verified.
- *Communication* relates everyday language to mathematical language and symbols, and the converse.
- *Connections* between mathematical concepts and between mathematics and other disciplines and everyday situations make mathematics meaningful to students. Such connections link perceptual and procedural knowledge and help students to see mathematics as an integrated whole.

INTEGRATING CONCEPTS AND PROCESSES IN THE CURRICULUM

Throughout the curriculum, the use of manipulatives and other models reinforces both the students' understanding and the variety of their learning styles. Several such models are included in the following discussion, whose objective is to illustrate how exploration of content and process can be integrated into the study of multiples throughout the curriculum. The concepts and processes used in each example are identified by capital and lower-case letters:

Concepts	Processes
P: Patterns, relationships, functions	r: Reasoning
N: Number and numeration	p: Problem solving
C: Computation	c: Communication
L: Language and symbolism	n: Connections
T: Tables and graphs	

In the earliest grades, making piles of three objects each and recording their observations in a variety of ways helps students to connect counting, number and numeration, computation, language and symbolism, and organization of data to each other and to everyday experiences. The development proceeds from one-to-one correspondence, as illustrated in figure 2.4, to introduction to multiplication by organizing the data as a graph, as shown in figure 2.5.

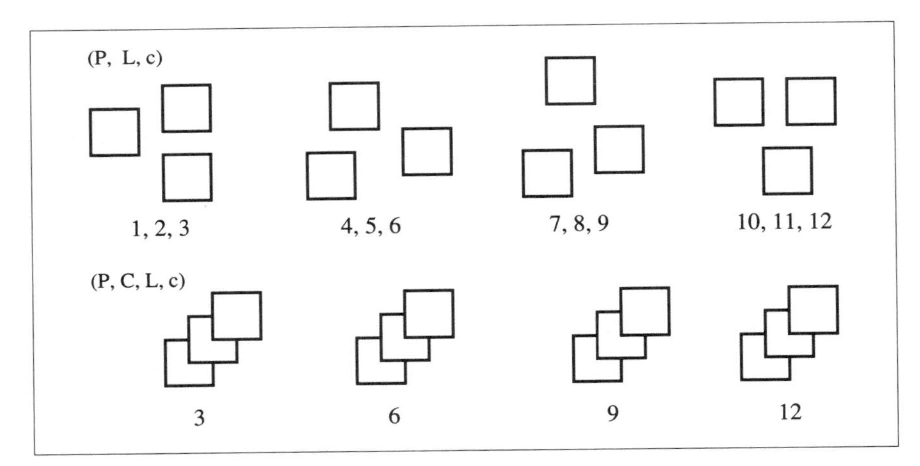

Fig. 2.4. One-to-one correspondence and skip counting

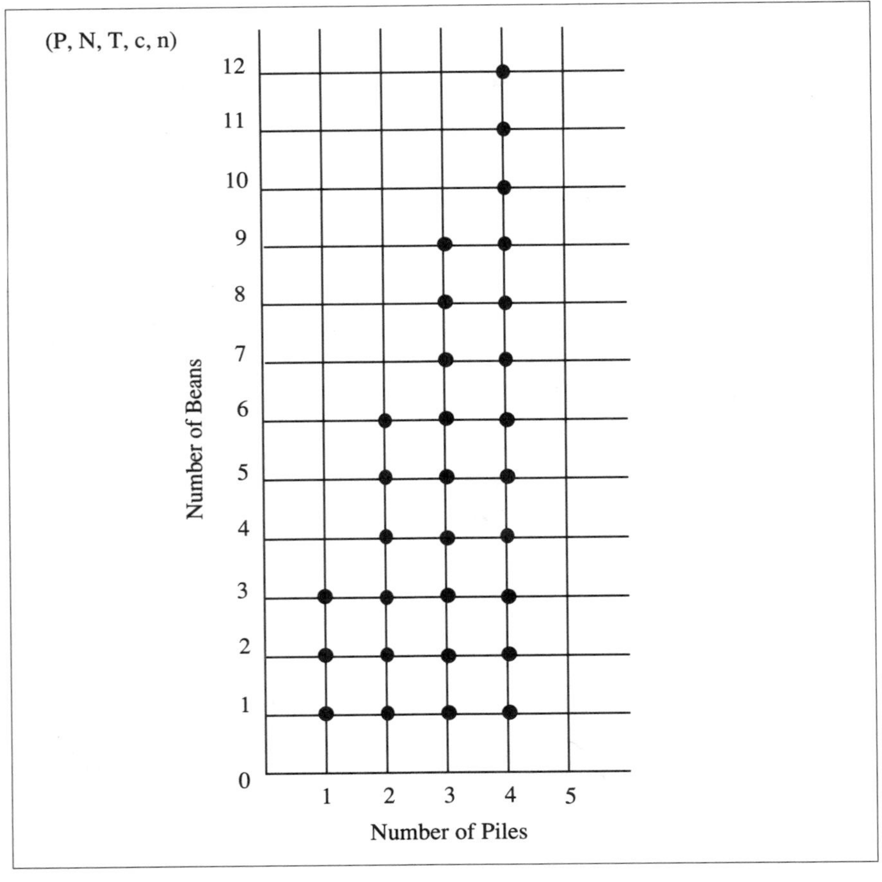

Fig. 2.5. Organizing data

Analyzing the graph introduces many mathematical concepts:

- Addition as an extension of counting (N, C, r)

$$\underbrace{1+1+1}_{3}+\underbrace{1+1+1}_{3}=6$$
$$3 + 3 = 6$$

- The commutative and associative properties (N, C, r)

$3+3+3=9$ $\quad$ $3+3+3=9$

$6+3=9$ $\quad$ $3+6=9$

- Introduction to multiplication (N, C, r)

Number of piles	1	2	3	4
Number of beans	3	3 + 3, or 6	3 + 3 + 3, or 9	3 + 3 + 3 + 3, or 12

The area model for multiplication is shown in figure 2.6.

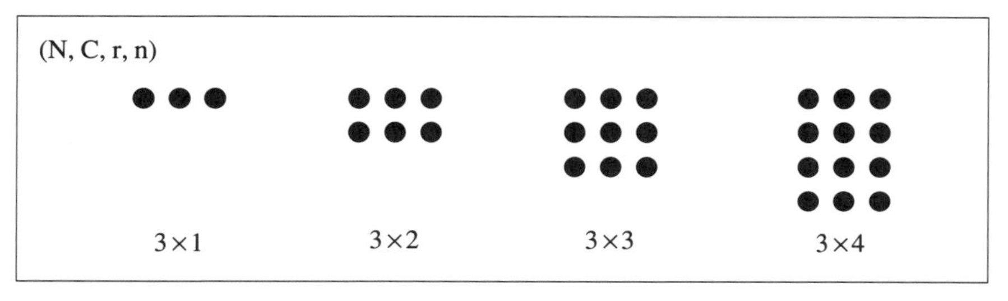

Fig. 2.6. Area model for multiplication

Introduction to variable units, fractions, and fraction notation is illustrated in figure 2.7. Making a measuring tape with each unit marked off in thirds, as in figure 2.8, relates the same concepts to length.

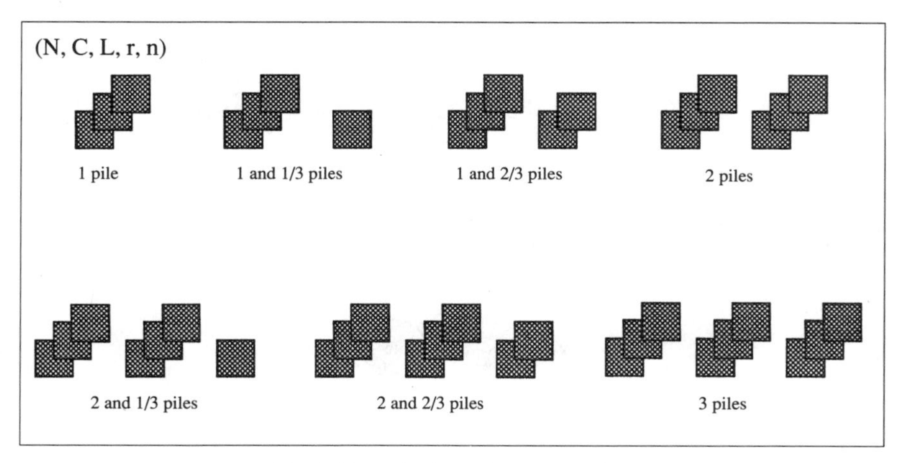

Fig. 2.7. Introduction to fraction notation

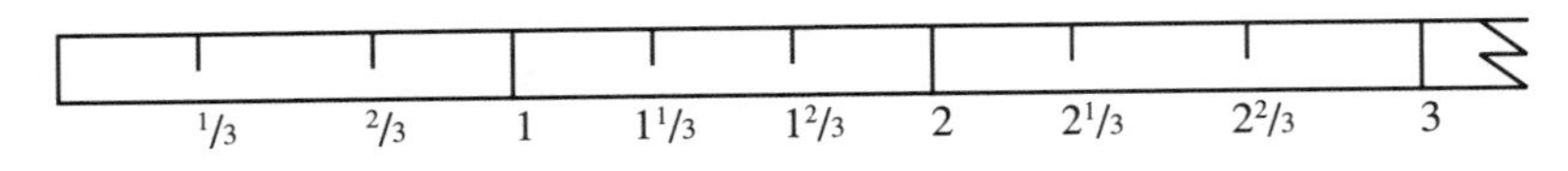

Fig. 2.8. Relating fractions to length

In the intermediate grades, these concepts can be extended by a study of multiples of 3 in a hundred chart such as shown in figure 2.9. The most obvious patterns concern the location of the multiples. (P, N, C, T, c):

1. Every third number is a multiple of 3.
2. The number of multiples in each row follows the pattern 3, 3, 4, 3, 3, 4, 3, 3, 4, ...
3. The multiples lie along diagonal lines.

Further examination reveals another interesting relationship:

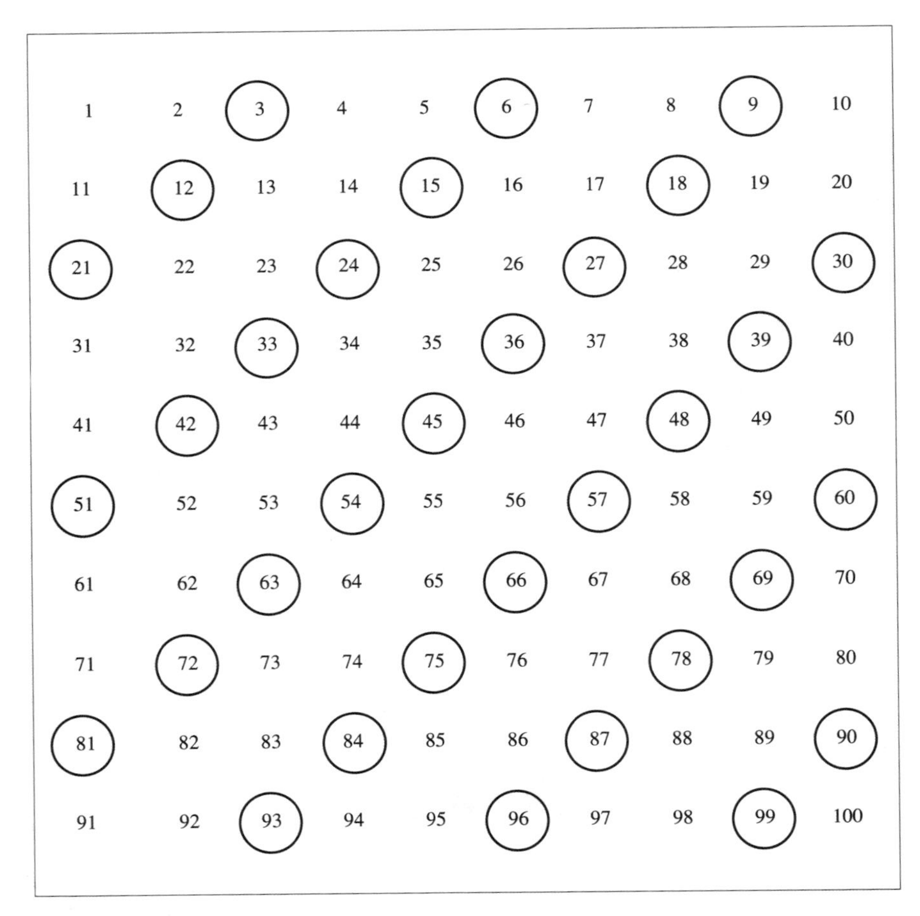

Fig. 2.9. Hundred chart

4. The sum of the digits in each multiple of 3 in a given diagonal has the same value—3, 6, or 9—according to the value of the multiple in the first row. (P, N, C, T, r, c)

Appropriate questions lead to the discovery of other relationships (P, N, C, T, r, p, c):

5. All the multiples of 3 that lie in the diagonal line beginning with 9 are multiples of 9, but only every other multiple of 3 that lies in the diagonal line beginning with 6 is a multiple of 6.
6. The numbers in the hundred chart can be rearranged to avoid breaking the diagonal lines on which the multiples of 3 lie. See figure 2.10.

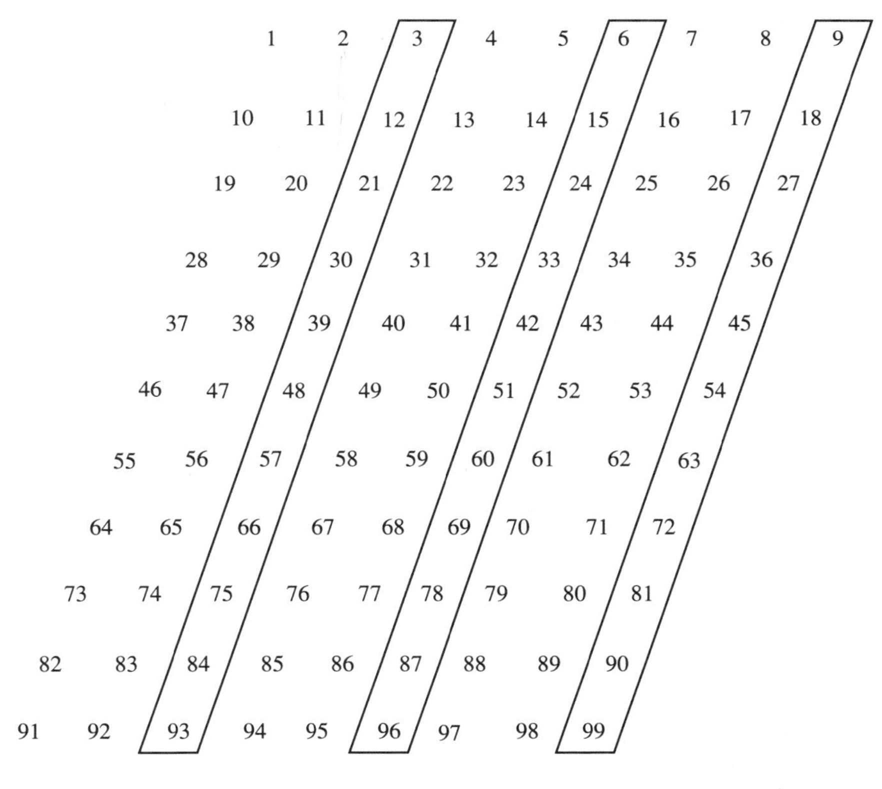

Fig. 2.10. Hundred chart with multiples of 3 on unbroken diagonal lines

7. The numbers in the hundred chart can be rearranged so that the multiples of 3 lie in columns, as in figure 2.11. In this arrangement, the sum of the digits of each of the numbers in the sixth column is 6. In fact, in each column, the sum of the digits of the numbers is equal to the number that heads the column. The

1	2	3	4	5	6	7	8	9
10	11	12	13	14	15	16	17	18
19	20	21	22	23	24	25	26	27
28	29	30	31	32	33	34	35	36
37	38	39	40	41	42	43	44	45
46	47	48	49	50	51	52	53	54
55	56	57	58	59	60	61	62	63
64	65	66	67	68	69	70	71	72
73	74	75	76	77	78	79	80	81
82	83	84	85	86	87	88	89	90
91	92	93	94	95	96	97	98	99
100								

Fig. 2.11. Hundred chart with multiples of 3 in columns

column headed by 9 is the most interesting. Notice that in addition to the fact that the sum of the digits is 9, the tens and ones digits in each succeeding entry from 9 to 90 increase and decrease by 1, respectively.

In the middle grades, all students should be able to explain why, in successive multiples of 9, the tens digit increases by 1 and the ones digit decreases by 1. All students should also be challenged to explain why the sum of the digits of the numbers in a given column is equal to the number that heads the column. A discussion of the various explanations that groups of students submit affords an excellent opportunity to reexamine the concept of place value. (P, N, C, T, r, p, c, n)

In addition, the relationships listed in the foregoing should be extended by graphing the functions $f(x) = 3x$, $f(x) = 6x$, and $f(x) = 9x$ in the coordinate plane. An examination of the graphs introduces the concepts of linear equations, slope, and such transformations as $(x, y) \rightarrow (x, y + b)$ and $(x, y) \rightarrow (x, ny)$. See figure 2.12.

Eratosthenes' sieve is a hundred chart on which the multiples of 2 through 10, starting with 2 times each of these numbers, are marked in different colors or with different symbols (fig. 2.13). It provides an excellent introduction to prime and composite numbers, prime factorization, and exponents. (P, N, C, L, T, r, c, n)

Further examination by the students of the multiples of 3, 6, and 9 relative to prime- and composite-number characteristics and an extension of their findings to multiples of other numbers lead students to discover divisibility rules and properties of terminating and repeating decimals. (P, N, C, L, T, r, p, c, n)

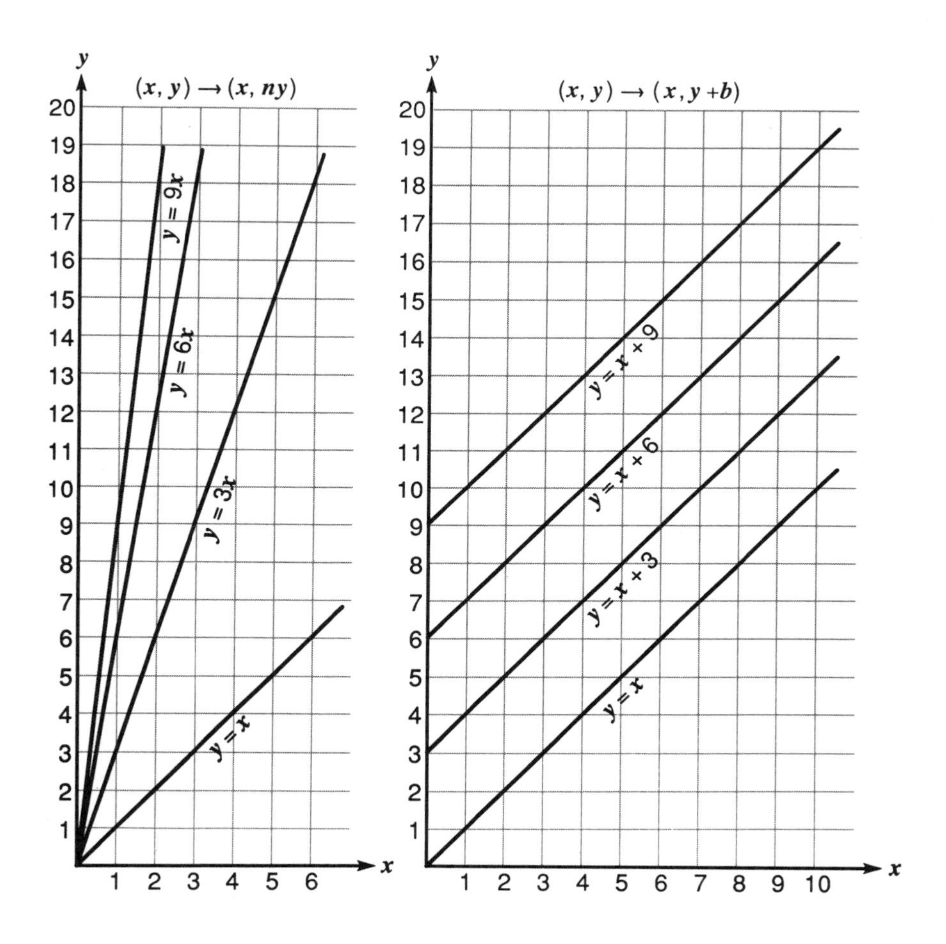

Fig. 2.12. Hundreds chart leads to exploration of linear eqations, slope, and transformations

SUMMARY

Preparation for success in algebra includes much more than computational proficiency. It means focusing on a broad range of mathematical content and processes so that students can become mathematically literate. Prior experiences must present opportunities for students to explore, conjecture, and reason logically so that skills can be acquired in ways that make sense to the students. These opportunities must focus on the development of understandings and on relationships among concepts and between the conceptual and procedural aspects of a problem.

Offering such experiences may require a rethinking of both the curriculum and the roles of teachers and students. Teachers must guide, listen, question, discuss, clarify, and create an environment in which students become active learners who explore, investigate, validate, discuss, represent, and construct mathematics.

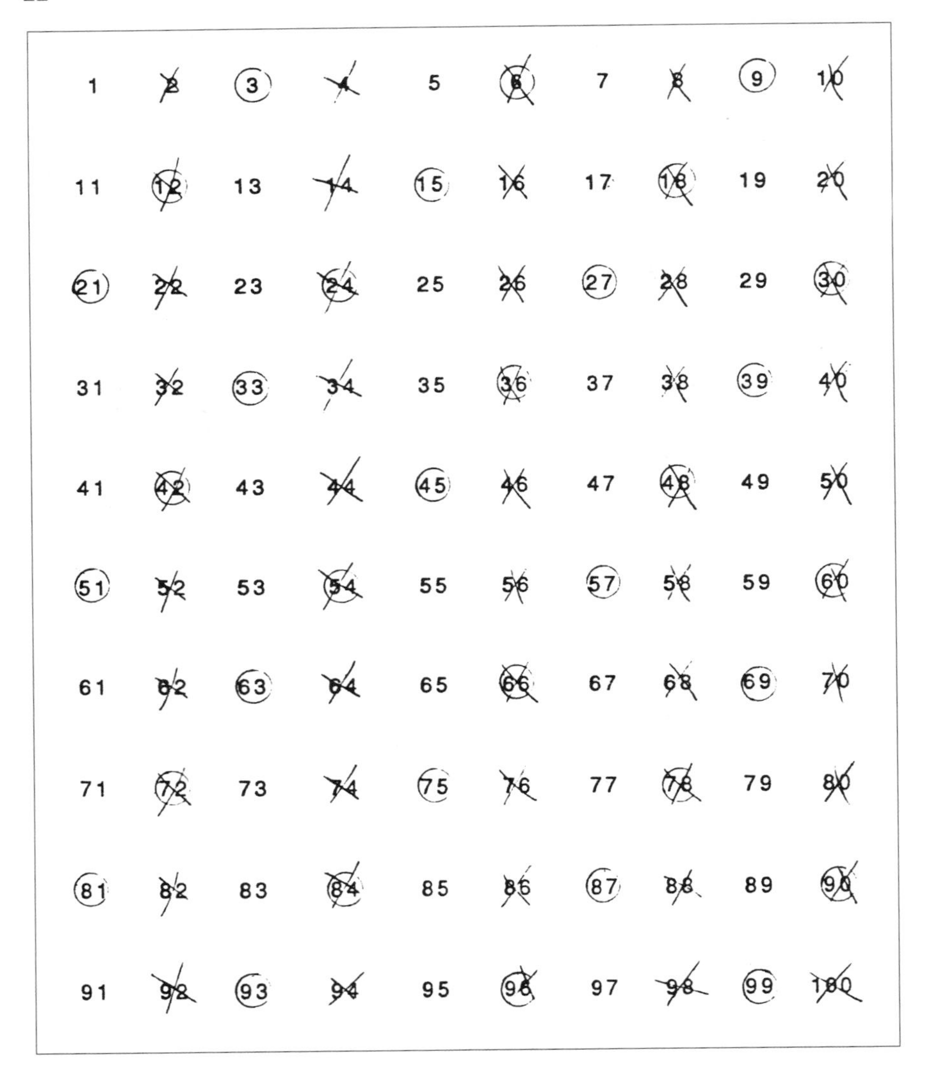

Fig. 2.13. Eratosthenes' sieve

BIBLIOGRAPHY

Bennett, Albert B., Jr. "Visual Thinking and Number Relationships." *Mathematics Teacher* 81 (April 1989): 267–72.

Blais, Donald M. "Constructivism—a Theoretical Revolution for Algebra." *Mathematics Teacher* 81 (November 1988): 624–31.

Burk, Donna, Allyn Snider, and Paula Symonds. *Box It or Bag It Mathematics*. Salem, Oreg.: Math Learning Center, 1988.

California State Department of Education. *Mathematics Framework for California Public Schools*. Sacramento, Calif.: The Department, 1985.

———. *Mathematics Model Curriculum Guide*. Sacramento, Calif.: The Department, 1985.

Chisko, Ann M., and Lynn K. Davis. "The Analytical Connection: Problem Solving across the Curriculum." *Mathematics Teacher* 79 (November 1986): 592–96.

Committee of Inquiry into the Teaching of Mathematics in the Schools. *Mathematics Counts (The Cockcroft Report)*. London: Her Majesty's Stationery Office, 1982.

Dossey, John A., Ina V. S. Mullis, Mary M. Lindquist, and Donald L. Chambers. *The Mathematics Report Card: Are We Measuring Up? Trends and Achievement Based on the 1986 National Assessment*. Princeton, N.J.: Educational Testing Service, 1988.

Driscoll, Mark. *Research within Reach: Secondary School Mathematics Teaching*. Reston, Va.: National Council of Teachers of Mathematics, 1983.

Gadanidis, George. "Problem Solving: The Third Dimension in Mathematics Teaching." *Mathematics Teacher 81* (January 1988): 16–21.

National Council of Teachers of Mathematics. *The Ideas of Algebra, K–12*. 1988 Yearbook of the National Council of Teachers of Mathematics. Reston, Va.: The Council, 1988.

National Council of Teachers of Mathematics, Commission on Standards for School Mathematics. *Curriculum and Evaluation Standards for School Mathematics*. Reston, Va.: The Council, 1987.

Ohio State Department of Education. *Eighth Grade Mathematics*. Columbus, Ohio: The Department. 1988.

______. *Teaching Mathematics: Elementary and Middle Grades*. Columbus, Ohio: The Department, 1988.

Usiskin, Zalman. "Why Elementary Algebra Can, Should, and Must Be an Eighth-Grade Course for Average Students." *Mathematics Teacher* 80 (September 1987): 428–38.

Wirtz, Robert W. *Mathematics for Everyone*. Washington, D.C.: Curriculum Development Associates, 1974.

Wisconsin Department of Public Instruction. *A Guide to Curriculum Planning in Mathematics*. Madison, Wis.: The Department, 1986.

3

THE TRANSITION FROM ARITHMETIC TO ALGEBRA

RICHARD D. LODHOLZ

ALGEBRA FOR EVERYONE! A worthy goal for mathematics education. The theme is certainly not new, and clearly it will provoke an argument from many teachers of mathematics in the secondary schools of the United States. An assumption of this publication is that the goal of "algebra for everyone" is both worthy and obtainable provided that we appropriately define just what comprises a desirable curriculum in algebra. Many high school teachers argue that more students could be successful if we reduce the demands in the traditional algebra course and, perhaps, stretch the content over a period of two or even three years. Such a shallow approach is not the intent of this book. We are assuming the established direction discussed by House in the 1988 Yearbook of the National Council of Teachers of Mathematics, *The Ideas of Algebra, K–12,* and outlined in *Curriculum and Evaluation Standards for School Mathematics* (hereafter called *Standards*), presented by NCTM in 1989. The purpose of this chapter is to address a key component for students' success in algebra: the mathematics curriculum prior to algebra, in the middle school years.

Immediately, a problem arises because of our traditional management of students and curriculum by courses. We usually think of algebra as a course, compartmentalized in a sequence of traditional courses. Worse, we think of the preparation for algebra as a course, or a sequence of courses, called prealgebra. Even the title of this chapter implies that something exists between arithmetic and algebra, some content bridging a gap between the arithmetic of the elementary school and the junior high school or high school course in algebra. Such a course mentality has caused the placement of students into distinct categories determined by their success with a skill-oriented program of arithmetic in the elementary grades. The Second International Assessment of Mathematics (Travers 1985) identified four tracks of eighth graders. If we push back this four-track separation to grade 5, we find at least three levels of mathematics content. First we find those students who master the traditional curriculum, taught under a behavioral learning psychology, and these students proceed to a course called prealgebra in grade 7. Next we find students not quite as successful, and they spend the middle grades reviewing some arithmetic in more complex exercises while they wait to enroll in algebra in the ninth grade. Finally, we observe the unsuccessful students, who stay in school and are relegated to a complete review of arithmetic. Typically, these students will never enroll in a course called algebra. Such a management scheme based on previous achievement in arithmetic skills and the organization of content based on disjoint courses is academically indefensible and is mentioned here to establish a premise for this chapter: just as algebra must be more than a disparate course in the curriculum, prealgebra must not be a single entity but rather a collection of knowledge, skills, and dispositions prerequisite for understanding algebraic concepts. Just as we do not have a course called pregeometry, but rather a strand of geometric concepts and skills, so should it be with prealgebra.

Although transition from arithmetic to algebra is philosophically defined by the NCTM's curriculum standards for grades 5–8 (NCTM 1989), it is, in reality, determined by the organization of our schools and the traditions of our schooling. American schools are organized as middle schools or junior high schools for this transition period. Although the *Standards* outlines a curriculum for all students and although ideally we strive to accomplish such a program, we confront the fact that not all students learn the same mathematics at the same pace and with the same understanding. If we view this middle-grades period as what Lynn Arthur Steen calls a "critical filter" (Lodholz 1986) to help organize subsequent high school study, we have a solid picture of the curriculum in the transition from arithmetic to algebra. For each of the three categories of students, the curriculum is, as the *Standards* states, basically the same. The difference is in the pacing and instructional style required for success. Attention to the content, pacing, and instructional methods during these transitional middle grades are, then, key components in the plan of "algebra for everyone."

The encouraging aspect is that the needed modifications in the present curriculum are within reach. We can, in fact, hold to some sorting of students by their talents, values, and interests if we change the emphasis in content and organize the pacing and instruction. Students in each of the three categories previously mentioned would be capable of understanding algebraic concepts at least by grade 10. The data in the recent international assessment (Travers 1985) indicates that about 10 percent of the students in the United States enroll in algebra by grade 8 and that about 65 percent are in a regular track in algebra by grade 9. The concern for guaranteeing success in algebra for all students is then directed toward the lower one-fourth of the student population, who presently do not even think about algebra. Much has been written in recent years (e.g., McKnight et al. [1987]) about the wasted mathematics curriculum during the middle school grades. Attention to those recommendations for content will eliminate the meaningless repetition of topics for the students unsuccessful in arithmetic and will help put them on the road to algebra.

But do we truly believe that everyone should take algebra? We need to think about the answer to this question before we proceed. Are students' needs different today from those in previous times? Yes and no. As House (1988) points out, two major forces operate on the content, instruction, and use of algebra in today's society—computing technology and social forces. The computing technology is a recent force to strengthen the argument, but mathematics educators have been concerned for years that algebra be within reach of all students. The NCTM president in 1932, John P. Everett, described algebra as primarily a method of thinking and presented the position that "thought, thinking processes, and the ability to appreciate mental and spiritual accomplishments are looked upon today as the rightful possessions of every individual" (Reeve 1932). Thus, the effort is not new, and the rationale for the effort is well documented. Algebra is a critical discriminator in this country for a student's future. It is crucial that all students have an opportunity for success in algebra.

As discussed in other chapters of this publication, a basic premise for accepting that everyone should succeed in understanding algebraic concepts and complete a course in algebra is the belief that algebra is more than memorizing rules for manipulating symbols and solving prescribed types of problems. Algebra is part of the reasoning process, a problem-solving strategy, and a key to thinking mathematically and to communicating with mathematics. Assuming some changes in the algebra course itself, what can we do in the transition years to guarantee students' success in completing algebra? We are challenging tradition in the management of students and curriculum and in the perception of mathematics education by the public and even by teachers of mathematics. However, the goal is realistic. Under the premise just stated, consider some of the reasons why algebra is a challenging subject for many students. If we address these trouble spots for all students, in general, and for students in the lower track, in particular, we have a solid plan for our effort to make algebra accessible to everyone. The key prerequisites for success in algebra are these:

- Understanding the technical language of algebra
- Understanding the concepts of variable
- Understanding the concepts of relations and functions

Content

The topics in the middle grades are well defined by the NCTM's *Standards* (1989). It is not a purpose of this chapter to restate that content, other than to endorse it heartily as meaningful for the preparation for algebra. The content discussed here is relevant to the key prerequisites stated in the foregoing. The focus on language development is so great that attention to language provides enhanced understanding for each of the three stated prerequisites.

One of the major reasons that students today do not succeed in algebra is that they do not correctly interpret the technical language of mathematics. Attention to language has numerous implications for both content and instruction. Although being attentive to language is a broad and perhaps vague directive, the basic routine for organizing both content and instruction should be moving from the descriptive language of the student to the more technical language of mathematics. We should think about language as (1) highlighting typical misconceptions; (2) discussing topics orally; (3) posing and composing problems; (4) writing conjectures, summaries, conclusions, and predictions; and (5) using symbols as a language.

Highlighting typical misconceptions. As summarized by Lochhead (1988), recent research indicates that a major part of the trouble students have in dealing with word problems is in the translation from the written language to the mathematical language. Students typically are given some practice with direct translation in mechanical problems. They even get practice with such exercises as writing an open sentence to restate the phrase "5 more than 3 times a number" as "$3x + 5$." However, this type of practice is usually isolated and out of context with applied problem situations. It becomes a skill in isolation and may even later cause difficulty with interpretations of meaningful sentences. The often used example of "there are six times as many students as professors" being written as "$6S = P$" gives much information as to students' misconception about the translation from written language to technical language.

How can we help? Students should be required to explain some of the typical conflicts between the language of arithmetic, with which they are familiar, and the more technical language of algebra, which they will need to master. In algebra we see that: $a \times b$ means the same as ab, but in arithmetic, $3 \times 5 \neq 35$; and $ab = ba$, but $35 \neq 53$. In arithmetic we find that $7 + ½ = 7½$ and $4 + 0.75 = 4.75$, but in algebra $2a + b$ does not mean $2ab$. Students should explain why not. If the sources of difficulty are misconceptions between written language and algebraic language, then the students should be confronted with these trouble spots prior to algebra.

Students should be required to write descriptive statements for such relations as $S/6 = P$, $S + P = 6$, $S = 6P$, $P = 6S$, $6S/P = T$, and $6S + P = T$. In the transition grades, students should struggle with the confusion between the different systems of representation. The trouble caused by the routine translation of the left-to-right matching of words and variables could be addressed by requiring students to describe the multiple arrangements of the same symbols, like those just presented. Attention in the problem sets of lessons should be given to providing practice with the translation process, highlighting the typical misconceptions, and forcing a struggle with the confusion. For example, writing about how the word *product* is used differently in social studies and in mathematics strengthens the understanding of its mathematical use.

For teachers of mathematics prior to algebra, it is a manageable task to require students to translate written language into proper symbolic statements of mathematics. The fact that present textbooks do not emphasize such exercises is irrelevant because teachers can simply compose them on a consistent and regular schedule. The only concern would be to make them of interest and make certain that students deal with the confusion caused by the translations. For example, consider these two written statements: (*a*) the number of males is two times the number of females; and (*b*) there were twice as many males as females. Once students make the translation $M = 2F$, they should be required to test the statement with examples that fit the criteria.

Discussing problems orally. Not much discussion of mathematics takes place in the classrooms in this country. Many teachers do not see a need for much discussion of mathematics by students because of the view of mathematics that they probably hold. The common conceptualization of mathematics as the quick attainment of an exact answer by some acquired routine conflicts with a desire for discussion. The content usually demands product questions, which do not require discussion, rather than process questions.

Discussion is crucial for motivating a desire to learn about a topic or to pursue a solution to a problem. As Sobel and Maletsky tell us (1988), mainly for this reason it is important to generate sufficient discussion about a problem in advance of finding a solution. Consistently, classes of students are not motivated to solve a problem. If the students are not interested, little value is realized in proceeding with an explanation of a solution. Predictions, guesses, conjectures, and confusion can each lead to discussions and defense of positions on processes and solutions. The content requires discussion of the processes involved and the various ways to solve the problem.

Discussion gives students a means of articulating aspects of a situation, which, according to Pimm (1987), helps the speaker to clarify thoughts and meanings. Discussion leads to greater understanding. Verbalizing externalizes the students' thoughts, makes them public, and provides the teacher with an invaluable tool for assessing students' understanding of the concepts. Verbalizing emphasizes attention to argument and develops the process of defending convictions. Verbalizing helps develop technical understanding because the descriptive talk and explanations must be worked toward communicating with mathematical terms and symbols.

Posing and composing problems. Implications for content in the transition period to algebra under this category are limited only by our ability to create variations on the initial problem situations. Silver and Kilpatrick (1987) relate that the problem variations should be progressive. After the students have solved a problem, we could change the context of the problem and pose it again. Next, we could change the data in the problem. A few lessons later we could use the technique of reversibility by giving the result and asking for the given portions of the problem situation. Also, we could make the problems more complicated by requiring multiple operations, extraneous data, and insufficient data.

Word problems should not be grouped as to type or style, but they should be organized more in line with the process for solution. For example, problems like finding how many sets of six whatevers are contained in seventy-two items is the same process as determining the rate of speed on a bicycle for traveling seventy-two miles in six hours. Requiring students to compose their own problems when given specific criteria and limiting information helps students to understand the process. Practice with posing similar and more complicated problems from given textbook problems addresses the progressive variations mentioned in the preceding paragraph.

Writing conjectures, summaries, predictions, and conclusions . Requiring the students to write or present oral conjectures, summaries, predictions, generalizations, and such, from collections of patterns, lists of data, or presentations of information is at the heart of understanding mathematics. The prevalent language of the teacher is what Pimm (1987) calls "surface mathematics language." Teachers currently train students to "cross multiply," to "take to the other side and add," to multiply by 100 by "adding two zeroes," and to "do to the top as you do to the bottom" when working with fractions. Instead, students should be required to draw their own conclusions about rules and should be permitted, at first, to derive their own algorithms. It should be our job in the classroom gradually to refine the descriptive everyday talk and explanations into the efficient, technical language and routines of mathematics.

How can students in the middle grades get too much practice with interpreting and determining patterns? How can we spend too much time requiring students to describe or explain patterns generated by using mathematics? It is practically impossible. Steen (1988) has presented the position that mathematics is recognized today as the "science of patterns." Data, analysis, deduction, and observation are schemes that present unlimited opportunities for students' discussion, writing, and explanation.

Mathematics as a symbolic system. Students must learn to use symbols as a language in which they can express their own ideas before they get to algebra. Then algebra will not be just a meaningless collection of rules and procedures. As Pimm (1987, p. 22) states, the meta-language of arithmetic is algebra. Most of the "laws of arithmetic" are taught explicitly in the meta-language. This condition and the fact that students do not understand the symbolic system cause problems and ambiguity. Usiskin (1988) presents many uses of the idea of "variable" that lead to different conceptions of algebra, and we see more clearly the importance of the language of mathematics and the importance of interpreting the symbolic system.

We would be hard pressed to find a better guide for giving students practice with the symbolic system than that given by Usiskin (1988). Understanding the concept of variable as a formula means that the students must have experience with manipulating numbers and symbols and with substituting values. Contrast that interpretation with the use of a variable in an "open sentence" like $17+x=35$, where the important idea is not the substitution but rather the relationship among the symbols. A third meaning comes from generalized statements like "$a+b=b+a$" in which variables are used to define properties for the operations over the numbers used in arithmetic. Yet another interpretation of variable relates to true variability, as given by relationships derived from data like "$y=2x+1$," which is more in line with a desired high school algebra course. Each of these understandings of variable requires appropriate language experience in the middle-school years, and students must be required to translate and generalize, using symbolism as a language to express their descriptive and numerical explanations.

Instruction

Educators should understand that the content of the mathematics curriculum and the instructional methods impact on each other because the content indicated in the preceding section dictates an instructional style that requires students to do, think, discuss, and interact. Appropriate instructional methods demand content that encourages such interaction. However, we have learned so much in the past fifteen years about how students learn and about teaching styles and classroom structures that we must pay special attention to recommendations for instruction in the transition years.

Skemp (1987) paints a clear picture of the desired instruction by relating the two views of "understanding" outlined by Stieg Mellin-Olsen of Bergen University:

"instrumental understanding" and "relational understanding." Instrumental understanding is categorized as rote learning with little need for explanations. It is basically a collection of rules and routines without reason. It is important because this type of understanding is the goal for most students and teachers in United States classrooms (Skemp 1987). Relational understanding is the goal proclaimed by the recent NCTM *Standards* and the method recommended by current research (NCTM 1989). A clear consensus emerges from the mathematics research and professional teaching and supervising organizations that we should work toward relational understanding, both in algebra and in the preparation for algebra. The claim of this chapter is that such focus will make attainable the goal of algebra for everyone.

It is true that under the present system that advocates instrumental learning, some students find success. It is also true that not all students can be successful under the present system. Not all students can recall, memorize, and maintain the huge collection of rules and routines necessary for instrumental understanding of algebra. Discussing conclusions and implications of current research in mathematics education, Peterson (in Grouws and Cooney [1988]) claims that the challenge for educators in the next decade will be to improve students' learning of higher-order skills in mathematics. Recent research and theory suggest that the following classroom processes might facilitate that relational understanding of mathematics:

(*a*) Focus on meaning and understanding.

(*b*) Encourage students' autonomy, independence, self-direction, and persistence in learning.

(*c*) Teach higher-order processes and strategies.

The findings from naturalistic studies (Grouws and Cooney 1988) of classrooms suggest that teachers do not currently emphasize these processes.

Although the Cockcroft Report (Committee of Inquiry into the Teaching of Mathematics in the Schools 1982) pertained to education in the United Kingdom, the wealth of information in that paper gives us ample suggestions for instructional routines. Mathematics teaching at all levels should include opportunities for—

- exposition by the teacher,
- discussion between teacher and pupils and among pupils,
- appropriate practical work
- consolidation and practice of fundamental skills and routines,
- problem solving, including the application of mathematics to everyday situations, and
- investigative work.

Hoyles (in Grouws and Cooney [1988]) draws from that report to give us specific recommendations for classroom routines that are pertinent for middle school youngsters. First, use mathematics in situations in which its power is appreciated. Then, reflect on the procedures used by various individuals and groups. Finally, attempt to apply these derived procedures to known efficient ones and to new domains and then to make appropriate connections.

We must change the dominating current method of instruction, which reaches for instrumental understanding. Students must be given opportunities to derive rules, make conjectures, and determine patterns. These opportunities come less from teacher direction and more from independent and small-group work. The task of the teacher then becomes relating the derived findings and rules to the appropriate language and algorithms of mathematics. We must teach our students to look for generalizations. Derived formulas mean a greater investment of time, but they require less drill and practice and are more often maintained. It is a longer fight, but a greater pay-off is realized. We must challenge students in the classroom to formulate principles and concepts for themselves.

Grattan-Guinness (1987) claims that "foundations are things we dig down to, rather than up from." He is stating that we must first produce some mathematics in a variety of ways and contexts before we try to systematize it. In American education, we usually attempt the foundations-up approach, which does not permit investigation, discussion, questioning, and conjecturing by students. This simple quote should tell us much about how to teach mathematics.

Pacing

Is it a compromise to talk about all students getting to algebra, but at different times? Do we weaken the effort of the *Standards* by saying, "Yes we believe all students should take algebra, but some in the eighth grade and some in the tenth grade?" This chapter argues that we do not. We surely do not believe that all students can obtain the same level of understanding of all mathematics. The strong argument presented here is that all students should complete, with success, the content of algebra. Some students will understand with greater meaning than others, and some students will need more time for the preparation for success in algebra. The claim is that consideration of pacing is crucial to reaching that goal.

Consider the previously mentioned levels, or categories, of students determined by achievement ranges from the first five grades of elementary school. Although we could do much to improve their program, the two top groups are not the key concern because they do have the opportunity to take algebra,which can open doors for future education and provide opportunities in the work force. As discussed at the beginning of this chapter, the target group is that lower one-fourth of the present student population who do not succeed in algebra. The content has been defined and instruction schemes suggested, and each of these is probably also acceptable to us, as teachers, for the two top categories. The implications for the lower group are overwhelming. We typically are concerned with managing these students, for whom it is easier to stress instrumental learning and dispensary methods of teaching. We must take the time to implement the recommended curriculum with these students, who require our most expert teaching.

If the content and instruction require student activities, thinking, discussion, and explanation first, then the argument is even more acute for these students at risk of not taking algebra. Recall that the format is for students to derive results and explain what they mean using their own descriptive language, and it is the task of the teacher to translate that effort into the desired technical language and concepts of mathematics. Successful students can usually meet lesson objectives within one or two classes under the present structure. Less successful students simply require more time. Some students can master the content suggested between arithmetic and algebra in one year. These students already take algebra in grade 8. Other students, the majority according to the Second International Assessment, take the same content over a two-year period. The claim in this chapter is that the third group, which traditionally has been relegated to a repetition of arithmetic with no expectation of enrolling in algebra, could develop relational understanding of the same content over a three-year period.

Such a scheme is not in conflict with the *Standards* and is realistic under the present management of students and curriculum. As John Everett told us in 1932, it would be most logical to limit success in algebra to a few, but such a scheme is abhorrent to our desires. If algebra is a way of solving problems, a plan for organizing data and information, part of a reasoning process, and a key component to thinking mathematically, then every individual should have the chance to succeed in algebra. Perhaps, if we change the emphasis in content and instruction and adjust the pacing, as described in this chapter, for students in the middle grades, then we will reach our goal of "algebra for everyone."

REFERENCES

Committee of Inquiry into the Teaching of Mathematics in the Schools. *Mathematics Counts (The Cockcroft Report)*. London: Her Majesty's Stationery Office, 1982.

House, Peggy A. "Reshaping School Algebra: Why and How?" In *The Ideas of Algebra, K–12 , 1988* Yearbook of the National Council of Teachers of Mathematics, pp. 1–7. Reston, Va.: The Council, 1988.

Grattan-Guinness, Ivar, ed., "History in Mathematics Education." Proceedings of a workshop held in Toronto, Canada, 1983. Cahiers D'histoire and de philosophie des sciences, no.21. Paris: Belin, 1987.

Grouws, Douglas A., and Thomas J. Cooney. *Perspectives on Research on Effective Mathematics Teaching*. Hillsdale, N.J.: Lawrence Erlbaum Associates and National Council of Teachers of Mathematics, 1988.

Lochhead, Jack, and José P. Mestre. "From Words to Algebra; Mending Misconceptions." In *The Ideas of Algebra, K–12*, 1988 Yearbook of the National Council of Teachers of Mathematics, pp. 127–35. Reston, Va.: The Council, 1988.

Lodholz, Richard, ed., *A Change in Emphasis*. Position papers of the Parkway Mathematics Project. Chesterfield, Mo.: Parkway School District, 1986.

McKnight, Curtis C., F. Joe Crosswhite, John A. Dossey, Edward Kifer, Jane O. Swafford, Kenneth J. Travers, and Thomas J. Cooney. *The Underachieving Curriculum: Assessing U.S. School Mathematics from an International Perspective*. Champaign, Ill.: Stipes Publishing Co., 1987.

National Council of Teachers of Mathematics, Commission on Standards for School Mathematics. *Curriculum and Evaluation Standards for School Mathematics*. Reston, Va.: The Council, 1989.

Pimm, David. *Speaking Mathematically: Communication in Mathematics Classrooms*. London, England: Routledge and Kegan Paul, 1987.

Reeve, W. D. , ed., *The Teaching of Algebra*. 1932 Yearbook of the National Council of Teachers of Mathematics. Washington, D.C.: The Council, 1932.

Silver, Edward A., and Jeremy Kilpatrick. "Testing Mathematical Problem Solving." Paper from Dialogue on Alternative Modes of Assessment for the Future. Mathematical Sciences Education Board, 1987.

Skemp, Richard R. *The Psychology of Learning Mathematics*. Hillsdale, N.J.: Lawrence Erlbaum Associates, 1987.

Sobel, Max A. , and Evan M. Maletsky. *Teaching Mathematics*. Englewood Cliffs, N.J.: Prentice-Hall, 1988.

Steen, Lynn Arthur. "The Science of Patterns." *Science,* 29 April 1988, pp. 611–16.

Travers, Kenneth J. , ed., *Second Study of Mathematics: Summary Report–United States*. Champaign: University of Illinois, 1985.

Usiskin, Zalman. "Conceptions of School Algebra and Uses of Variables." In *The Ideas of Algebra, K–12,* 1988 Yearbook of the National Council of Teachers of Mathematics, pp.8–19. Reston, Va: The Council, 1988.

4

ENHANCING THE MAINTENANCE OF SKILLS

DAVID J. GLATZER
GLENDA LAPPAN

The NCTM's *Curriculum and Evaluation Standards for School Mathematics* (*Standards*) (1989) presents a vision of mathematical power for students that forces us to rethink in fundamental ways what is important for students to know and be able to do in mathematics. The title of this chapter,"Enhancing the Maintenance of Skills" (in algebra), might in the past have conjured up an image of "skills" that could be characterized as algebraic manipulation of symbols or solving word problems of predictable types. The picture presented in the previous chapters of *Algebra for Everyone* describes algebraic "skills" much more broadly. The central core of being algebraically skillful is being able to confront a problem that might be somewhat vague and use mathematical thinking and reasoning to formulate a mathematical version of the problem; represent or model the mathematical problem in, perhaps, more than one way; solve the mathematical problem using whatever tools are appropriate; and interpret the solution in the context of the original problem. A final step in the process described might even be to formulate additional questions or generalize the solution strategies and results to other situations. Such a process is mathematically powerful in the spirit of the *Standards*. The skills associated with this vision move far beyond skill in manipulating symbols.

Broadening the complexity and interaction of skills has significant ramifications for both development and maintenance of skills. None of us would argue that becoming skillful at measuring the ingredients for a recipe is all there is to cooking. Nor could we generate support for endless practice at beating egg whites. We recognize that a skillful cook not only knows the concepts and procedures that help to make a cake but has a sense of how these things fit together and need to be adjusted on a particular day with a particular batch of ingredients and a particular oven to produce an excellent cake. The flexibility and confidence to abandon the algorithm (the recipe) and make adjustments to fit the current problem are what make a good cook. We should want nothing less for students of algebra than having the power and flexibility to confront situations and use the power of mathematics to construct ways to understand the situation better.

The enhancement and maintenance of skills in algebra will be discussed from two perspectives—

- situating algebra in contexts that give meaning to concepts and procedures and the connections among them; and
- developing and maintaining skills in the language, symbols, and syntax of algebra.

Examples are given to illustrate each of these perspectives. The intention is to raise questions and issues that should be considered in assigning homework and classroom activities to develop and maintain skills.

LEARNING AND PRACTICING IN CONTEXTS

One of the main messages of the NCTM's *Standards* (1989) is that mathematics should be learned in contexts that help students to see connections within mathematics, between mathematics and other school subjects, and between mathematics and its uses and applications in the real world.

The introduction to the *Standards* makes the following statement (1989, 9–10):

> Traditional teaching emphases on practice in manipulating expressions and practicing algorithms as a precursor to solving problems ignore the fact that knowledge often emerges from the problems. This suggests that instead of the expectation that skill in computation should precede word problems, experience with problems helps develop the ability to compute. Thus, present strategies for teaching may need to be reversed; knowledge often should emerge from experience with problems. In this way, students may recognize the need to apply a particular concept or procedure and have a strong conceptual basis for reconstructing their knowledge at a later time.

The arguments for the benefits of learning mathematics in contexts that motivate and give meaning to the ideas learned are equally valid for the maintenance of skills. Skills should also be practiced in situations that require thinking, that show connections, that stimulate questions, and that further deepen understanding.

The *Standards* offers a way for us to analyze the maintenance activities that we use. Questions such as the following can be used to guide our selection of tasks for assignments in developing and maintaining algebraic skills:

- Do the tasks posed require problem solving, reasoning, and communication of ideas and strategies (including communication in the language of mathematics —algebra)?
- Do the tasks posed help students to consider connections to previously learned mathematics or to explore ideas that will be the subject of upcoming study?
- Do the tasks posed show mathematical connections to other school subjects or to valuable applications of mathematics?
- Do the tasks posed push students to consider their emerging knowledge from a new perspective?
- Are developing concepts surrounded by many situations that deepen meaning?
- Are procedures developed and used flexibly in such a way that students see the

power of the creation of procedures and the need to judge the appropriateness or inappropriateness of a procedure in particular situations?

- Are the tasks posed engaging to students?
- Do the tasks posed require any reflection on, or interpretation, of results?
- Do the tasks posed keep students engaged with important concepts, skills, and procedures over time (distributed practice)?

The following sections show types of situations that have many of the desired characteristics implied by the questions raised.

Physical Investigations

Consider the following situation: a same-color staircase is made from Cuisenaire rods (see fig. 4.1.) Each time a rod is added to the staircase, it is offset by the space of a white (unit) rod. Suppose you are working with light green rods (3 units in length). Predict the volume and surface area (S.A.) that your staircase will have when you add the tenth rod,..., the twenty-fifth rod,..., the *n*th rod. Is the pattern the same if you build a staircase from rods of a different color (length)?

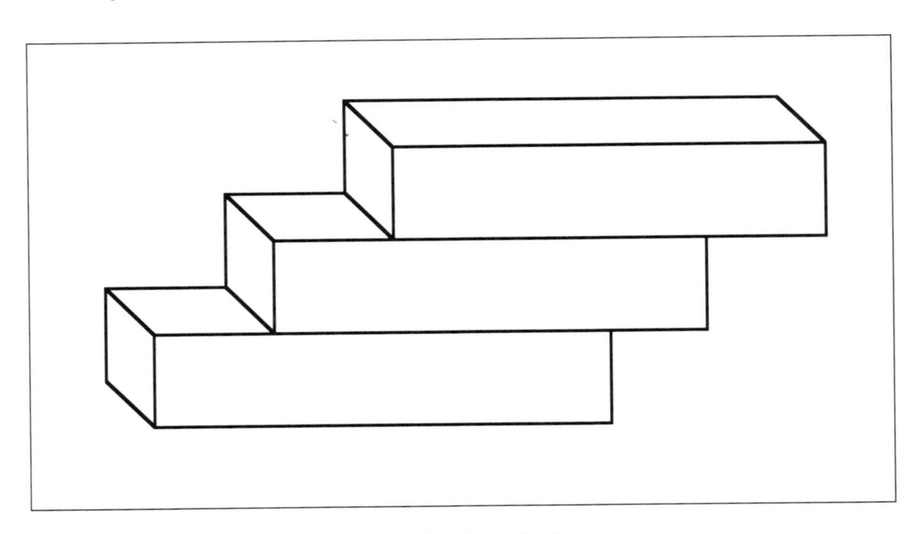

Fig. 4.1. Cuisenaire-rod staircase

Examples of data collected and organized for several different rods (fig. 4.2) show the richness of the situation. The pattern for volume is simply a multiple of the volume of the basic rod. The pattern for surface area is always an arithmetic sequence , but the rule is different for each basic rod. In each instance, variables are being used as a means of describing and generalizing a pattern in numbers. When we ask follow-up questions, we can probe other interpretations of the meaning of variable. For example, the question "What will the surface area be for a twenty-five-yellow-rod staircase?" asks the student to replace the variable in the expression with a member of the replacement set and to compute the value of the rule or function for that value of the independent variable. Next we can also ask problems of the sort, "Could the surface area ever be 625 for a staircase of yellow rods?" We are viewing the variable as standing for a specific unknown, namely, the solution to the equation $14n + 8 = 625$. If this solution gives a whole-number value of n, then the original physical problem posed has a solution, but if the solution to the mathematical equation is a fraction, the real problem has no solution.

Light Green Rods (3 Units in Length)

No. in Staircase	Volume	S.A.
1	3	14
2	6	24
3	9	34
4	12	44
.	.	.
.	.	.
.	.	.
n	$3n$	$10n + 4$

Purple Rods (4 Units in Length)

No. in Staircase	Volume	S.A.
1	4	18
2	8	30
3	12	42
4	16	54
.	.	.
.	.	.
.	.	.
n	$4n$	$12n + 6$

Yellow Rods (5 Units in Length)

No. in Staircase	Volume	S.A.
1	5	22
2	10	36
3	15	50
4	20	64
.	.	.
.	.	.
.	.	.
n	$5n$	$14n + 8$

Dark Green Rods (6 Units in Length)

No. in Staircase	Volume	S.A.
1	6	26
2	12	42
3	18	58
4	24	74
.	.	.
.	.	.
.	.	.
n	$6n$	$16n + 10$

Fig. 4.2. Data for staircase made from different rods.

Finally, the big question that raises the level of generalization is "Can we find a rule to predict the volume and surface area for a staircase n rods high built from any rod you choose?" In other words, does a super rule exist that lets you enter both the rod chosen and the number of rods in the stack to find the surface area or volume? The physical nature of this problem stimulates many ways of thinking and reasoning about the patterns observed. The situation provides a context against which mathematical conjectures can be tested for the sense that they make. Figure 4.3 presents two different ways of arriving at a super rule. The results lead to the problem of showing that the two different strings of symbols (the two rules) are mathematically equivalent.

Length	Surface Area for n Rods in Staircase
2	$2(4n + 1)$
3	$2(5n + 2)$
4	$2(6n + 3)$
.	.
.	.
.	.
m	$2[(m + 2)n + (m - 1)] = 2mn + 4n + 2m - 2$

No. in Staircase	Surface Area For Staircase Constructed With Rods of Length (Rods × Surface Area – Overlap)
2	$2(4m + 2) - 2(m - 1)$
3	$3(4m + 2) - 4(m - 1)$
4	$4(4m + 2) - 6(m - 1)$
.	.
.	.
.	.
n	$[n(4m + 2) - 2(n - 1)(m - 1)] = 2mn + 4n + 2m - 2$

Fig. 4.3

Within a single problem situation, students have practiced gathering and organizing data, looking for patterns, analyzing and generalizing patterns, and using algebraic language. They have considered three different uses or interpretations of the idea of variable, solved mathematical problems that must then be interpreted in a physical situation to see if the solutions make sense, and grappled with "different looking" algebraic statements or expressions to decide if they are mathematically equivalent—all in a situation that also reviews measuring surface area and volume.

Applications to Real-World Problems

Problems that allow the practice of skills in real-world contexts have the added advantage of showing the value of mathematics in our culture.

> The Bouck family is going on vacation. The Boucks are planning to fly to Denver, Colorado, and then rent a car for eight days to drive through the Rocky Mountains to Yellowstone National Park and back to Denver. Mr. Bouck asked his daughter, Emily, to help him figure out the best deal on rental cars. They can choose from three car-rental agencies, whose terms are shown below:

U-Can-Rent-It	Good Deal Car Rental	Tri-Harder Car Rental
Daily rates	Daily rates	Daily rates
Mid-size	Mid-size	Mid-size
\$38 per day	\$42 per day	\$48 per day

U-Can-Rent-It	Good Deal Car Rental	Tri-Harder Car Rental
75 miles free per day; \$0.32/mile over 75 Full tank gas	100 free miles per day; \$0.30/mile over 100 Full tank gas	Free unlimited mileage \$2 per day for 4-door car. \$12 one-time charge for fuel 1/2-tank gas

In addition, Mr. Bouck belongs to a travel club that gives him 10 percent off the total bill if he rents from "Good Deal" and 5 percent off if he rents from "U-Can-Rent-It" but nothing off from "Tri-Harder." From which company should Emily tell him to rent?

One very valuable aspect to this problem is that its solution depends on the particular description that each student gives for the Bouck family's trip. Do they drive straight to Yellowstone National Park and back? Do they explore along the way and around the Yellowstone Park area? How many miles per day do they average over the eight days? The students could also be asked to describe more generally how to figure the bill for each car-rental agency (write the function) and to describe under what circumstances, if any, each dealer has the best rates. Graphing the three functions gives a picture that allows quick visual comparisons.

Reverse the Directions

The following examples of problems that can be used to maintain skills while deepening understanding use reversing the directions as a context for exploration. Algebra is often taught by giving an example and showing a procedure for solution followed by practice of like problems. This process can be reversed in ways that often require more open-ended thought on the part of the students. The first illustration of reversing is to ask students to create examples of specific kinds.

Give an example of—

- a quadratic equation with +2 as a root;
- a quadratic equation whose graph has no x-intercepts;
- three different lines that pass through the point (–2, 1).

Another kind of reversing is to give the mathematical model and ask students to write stories to fit the model.

Write two different stories that would fit the mathematical model given:

- $f(x) = 2x + 5$
- $y \leq x^2 + 3x - 5$

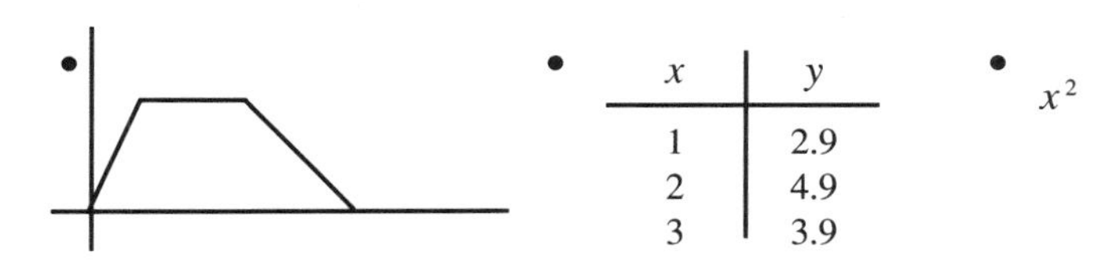

x	y
1	2.9
2	4.9
3	3.9

- x^2

At all grade levels in the NCTM's *Standards* the argument is made that an important mathematical goal is to learn to represent situations in different ways and to see the interaction of these different forms of representation. Developing skill at representing and interpreting representations is critical for the learning of algebra. We have seen, in the problems discussed, the interaction of physical and verbal descriptions and tabular, graphical, and symbolic representations. Graphical representation and its interplay with both symbolic and tabular information deserve special highlighting. With computers and graphing calculators, we have the tools to create graphs with virtually the push of a button. This power opens up the possibility of using graphs to help give meaning to algebraic expressions and functions, and vice versa.

An essential aspect of algebra is developing skill with algebra as a language with symbols and syntax that can be used to represent a situation, manipulate the representation, and yield new information about the situation. Students need meaningful practice activities to help their development of algebra as a language.

DEVELOPING AND MAINTAINING SKILLS IN THE LANGUAGE, SYMBOLS, AND SYNTAX OF ALGEBRA

Teachers need to offer practice experiences that go beyond the mechanical tasks usually associated with algebra as a language. The following is a list of aspects that teachers should consider in designing practice activities for developing and maintaining algebraic language skills:

1. Developing processes for structuring approaches to algebraic tasks
2. Creating exercises that highlight the critical attributes related to a particular skill or concept
3. Providing opportunities for students to verbalize the nature of the task and the type of answer expected
4. Offering opportunities for students to discuss and write responses to questions dealing with the key concepts being learned
5. Selecting exercises that anticipate skills and formats to be developed at a later stage
6. Designing exercises that integrate a number of ideas and require students to appreciate mathematics as a whole

Developing Processes

As students learn elementary algebra, they need to develop an appreciation for the power of procedures in algebra. A structured algebraic procedure can be used to solve a whole class of problems and, moreover, can become a record of one's thought and actions in solving the problem. This concept can be illustrated by the following examples:

a) Evaluate $3x^2y - 4xy^2$ for $x = 2$ and $y = -1$

Solution:

$$\begin{aligned} & 3x^2y - 4xy^2 \\ &= 3(2)^2(-1) - 4(2)(-1)^2 \\ &= 3(4)(-1) - 4(2)(1) \\ &= -12 - 8 \\ &= -20 \end{aligned}$$

b) Find the value of k such that the line passing through $(k, 5)$ and $(10, 8)$ has a slope of ¾.

Solution: slope = slope

$$\frac{y_2 - y_1}{x_2 - x_1} = \frac{3}{4}$$

$$\frac{8-5}{10-k} = \frac{3}{4}$$

$$\frac{3}{10-k} = \frac{3}{4}$$

$$10 - k = 4$$

$$6 = k$$

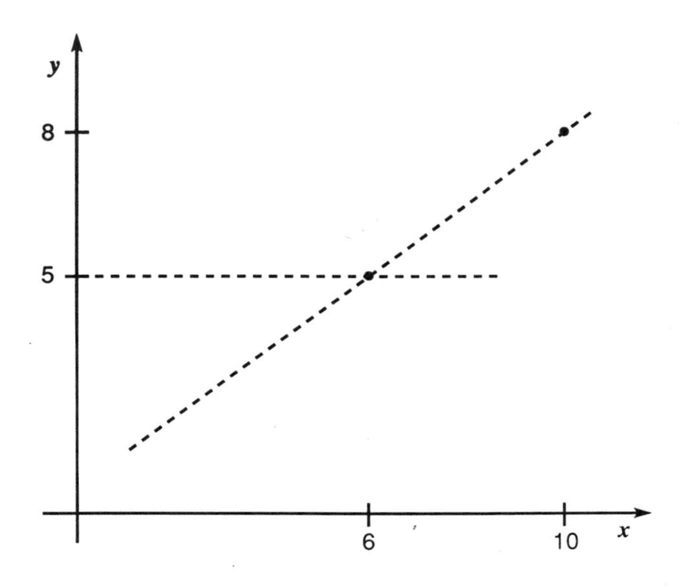

A sketch of the situation would be part of the expected structure of the solution strategy. The sketch should help the student determine the reasonableness of the solution.

Critical Attributes

Exercise sets can be carefully structured and sequenced so that an important aspect or attribute of a new skill or process can be highlighted. Consider the following sequence of exercises related to combining like terms:

a) $12x + 5x$	*g*) $12x + 5x + 6$
b) $12x + 6x$	*h*) $12x + 5x + 66$
c) $12x + 11x$	*i*) $12x - 3x + 6$
d) $12x - 7x$	*j*) $12x + 3(x + 2)$
e) $12x - 12x$	*k*) $12x + 3(x - 4)$
f) $12x + 5x + 6x$	*l*) $12x + 3(x + 1) + 2x$

In keeping the first term fixed at $12x$ the student is encouraged to concentrate on the critical attributes that change. In addition to furnishing practice in doing the problems, having students write about changes in exercises *a–l* helps students focus on such essentials as distinguishing terms and factors and at the same time gives the teacher information to assess students' understanding and plan future instruction.

Nature of the Task and the Type of Answer Expected

Students perform a great variety of mechanical tasks in algebra, and in so doing they often confuse the various processes. For example, in simplifying the expression $3x - 4 + 6x - 14$, mysteriously the student introduces a zero and comes up with $x = 2$ ($3x - 4 + 6x - 14 = 0$). Practice situations need to anticipate this potential confusion by requiring students to verbalize the nature of the task being performed and the type of answer associated with the task. Consider the following examples:

a) Simplify: $5(x + 2) - 3(2x + 4)$

Nature of the task: Simplifying an expression

Expected result: Another expression.

b) Solve for x: $3x + 7 = 4(x - 2)$

Nature of the task: Solving a linear equation

Expected result: A single value for x that makes the original equation true (assuming the original equation is a typical conditional equation)

c) Solve : $\begin{Bmatrix} x + 4y = 8 \\ 2x - y = 4 \end{Bmatrix}$

Nature of the task: Solving a system of two linear equations

Expected results: If the two lines intersect, a unique solution that consists of an ordered pair (x, y) that would make each equation true; if the lines are the same, an infinite set of solutions (x, y) that make each equation true; or if the lines are parallel, no solution.

Discuss and Write Responses

With an emphasis on mechanical tasks, students learning algebra often have difficulty in verbalizing responses to key questions concerning the major skills and concepts being learned. Students often respond, "I have the answer, but I don't know why it works" (or some similar statement). Teachers need to present daily opportunities in which the focus is on developing confidence in verbalizing the concepts. Students need to discuss and write responses. A small-group format can be effective for this type of activity. For example, the student can be asked, "How do you know that –7 is not the solution to the equation $3x + 4 = 6x - 2$?" A typical response might be, "When –7 is substituted for x, you do not get a true statement." Additional sample suggestions could include the following:

a) Can you tell whether the point (3, 4) is on the graph of the line $4x - y = 8$?

b) How do you know whether the line $y = 4x$ is parallel to the line passing through (0, 0) and (1, 8)?

c) Discuss whether the following is an identity: $|x + y| = |x| + |y|$

Anticipation

In formulating practice assignments, it is important for teachers to anticipate skills and formats that students will encounter later. For example, in working with slope, students will need to use the format

$$\frac{(y_2 - y_1)}{(x_2 - x_1)}.$$

Earlier in the school year, in evaluating expressions involving signed numbers, teachers can use the slope formula as a vehicle for some of the practice with signed numbers. A second example involves the quadratic formula. Under normal circumstances a student would not encounter the quadratic formula until the later stages of elementary algebra. However, it is possible to expose students to the kinds of expressions and computations involved by designing exercises with signed numbers and radicals, such as the following:

Simplify:

a) -2 ± 4

b) -8 ± 8

c) $\frac{-2 \pm 4}{2}$

d) $\frac{-4 \pm 8}{12}$

e) $-2 \pm \sqrt{9}$

f) $-2 \pm \sqrt{16}$

g) $\frac{10 \pm \sqrt{25}}{3}$

h) $\frac{-3 \pm \sqrt{8}}{2}$

INTEGRATING IDEAS

Practice situations need to provide for ongoing review, application, and integration of previously learned ideas. Mixed practice needs to include opportunities for students to transfer learning to new situations. Consider the following example:

In a right triangle, the hypotenuse and the long leg differ by 1 unit. Find an expression for the other leg.

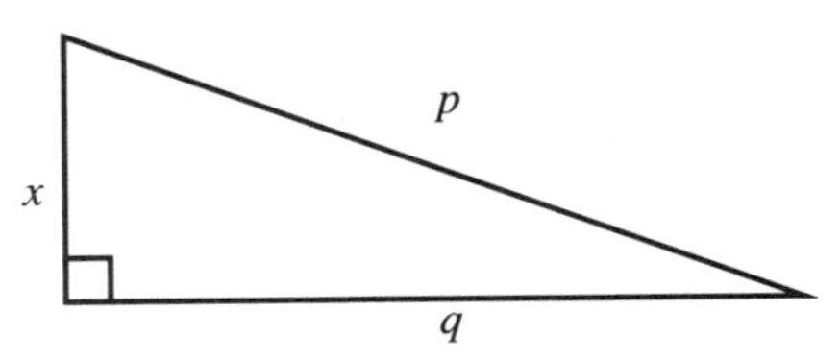

Solution:

Given:
$$p - q = 1$$
$$x^2 + q^2 = p^2$$
$$x^2 = p^2 - q^2$$
$$x = \sqrt{p^2 - q^2}$$
$$x = \sqrt{(p+q)(p-q)} = \sqrt{(p+q) \times 1} = \sqrt{p+q}$$

This problem could be used to integrate several skills and concepts learned earlier, such as factoring and the Pythagorean theorem. Students need opportunities to experiment with the use of algebraic skills and problem-solving strategies without always being told which to employ. Practice that uses applications can be a good arena for this added dimension.

Meaningful practice is essential for success in learning algebra. Students need to view practice as an integral part of the learning process. The major goal should not be to "finish the assignment" but rather to become mathematically powerful. Each assignment should offer students opportunities to verbalize their thoughts, strategies, and understandings and to develop confidence in applying algebraic skills and procedures to help make sense of many different kinds of situations.

REFERENCE

National Council of Teachers of Mathematics, Commission on Standards for School Mathematics. *Curriculum and Evaluation Standards for School Mathematics*. Reston, Va.: The Council, 1989.

5

TEACHER EXPECTATIONS OF STUDENTS ENROLLED IN AN ALGEBRA COURSE

ROSS TAYLOR

OF ALL THE EXPERIENCES a student has in kindergarten through twelfth grade, successful completion of the first course in algebra has the greatest impact on future opportunities. Algebra is the fork in the road where one direction leads to opportunity and the other to limited options for further education and promising careers. None of the other disciplines has a similar decision point. For example, in most high schools all students are required to take English every year. In science, the ability to succeed in chemistry or physics depends more on prior mathematics courses than on prior science courses. Social studies courses tend not to depend on knowledge from previous social studies courses.

Whereas some other countries have elaborate testing systems that determine which students will go on to higher education, in the United States and Canada success in precollege mathematics plays a major role in the sorting process. Mathematics has been described as the "critical filter" for determining a student's options (Sells 1978). The elementary algebra course is the most critical point of the critical filter.

In the typical precollege high school sequence (algebra 1, geometry, algebra 2, precalculus), each course successively screens out more students. Only a small number of students complete the full four-year mathematics sequence that prepares them for college majors in mathematics, science, engineering, and other technical fields. We can anticipate that in the information society of the future, a good background in mathematics will be required for most fields. Elementary algebra, geometry, advanced algebra, and additional work in computers, statistics, and probability will be necessary to open doors in practically every field.

Mathematics courses have tended to screen out certain groups, such as minorities, females, and students from low-income families, in excessive numbers. This process has contributed to the inequities in employment and income that exist in society today. To achieve an equitable society, we must change the algebra course from a filter that screens out segments of our population to a pump that propels all students toward opportunity (National Research Council 1990). In the process we can develop the brainpower needed for the society of the future.

Expectations and the Self-fulfilling Prophecy

To help students achieve success in algebra, we need to begin by believing that they can be successful. We can learn something from a look at the difference in expectations between Asian societies and our own. Whereas Asians tend to attribute success in mathematics to effort, we tend to attribute success to ability. (Stevenson et al. 1986.) Unfortunately, many in our society accept the notion that a large segment of our students do not have the ability to learn algebra. This idea leads to a self-fulfilling prophecy—with the result that many students do not successfully complete algebra—and, thereby, limits their opportunities for the future.

A major reason that students continue in mathematics or avoid it is their perception of how good they are at mathematics (Armstrong and Kahl 1979). If they believe that they have ability in mathematics, then they are likely to continue to study mathematics. Students who believe they lack mathematical ability tend to drop out of mathematics. Teachers should take every opportunity to give positive reinforcement to students. Whenever students are successful, they should receive the message that they were successful because they applied themselves and that if they continue to apply themselves, they can expect more success. Students who do poorly should not be allowed to come to the conclusion that they are dumb, that they can never learn mathematics, and, therefore, that they should drop out of mathematics. The difficult job of developing self-confidence for these students can be accomplished by teaching them how to learn mathematics and offering them motivating experiences in which they can have success.

The idea that it is all right not to be good in mathematics should not be acceptable in our society. Today people who specialize in mathematics often receive this reaction to their specialty from others at social occasions: "Oh, well, I was never very good at mathematics." No one ever says to a reading specialist that they were never very good at reading. To prepare for the future, we need to convey the expectation that everyone can learn mathematics and that avoiding mathematics is unacceptable.

The belief that virtually all students can learn algebra places a whole new perspective on teaching algebra. We cannot write off unsuccessful students as simply not having mathematical ability. Some possible reasons for lack of success in algebra are (1) a lack of prerequisite knowledge, (2) the failure to study properly, and (3) inadequate instruction.

Expectations and Prerequisite Knowledge

Before beginning a lesson, an algebra teacher should take steps to ensure that the students have the necessary prerequisite knowledge. Henry Luce, former publisher of *Time* and *Life* magazines, is reported to have told his editors, "Never overestimate what the readers know, and never underestimate what they can learn." We should replace "readers" with "students" and apply this adage to teaching algebra. If the teacher checks to see that the students have the necessary prerequisite knowledge, the students can be provided with the appropriate advance preparation for successful learning during the lesson.

Throughout the algebra course, teachers continually need to use a variety of

techniques to assess and review prior mathematical knowledge. This approach is especially important early in the course to ensure that students' first experience with algebra is successful. This success will contribute to high expectations, which will contribute to more success. Expectations illustrate the validity of the cliche "Nothing succeeds like success."

Expectations and Studying Properly

Initially, the teacher should establish clear expectations with respect to the classroom environment, the completion of homework, and the assumption of responsibility by students for their own learning. Students should begin with the assumption that they must apply themselves if they are to learn algebra. They should recognize that algebra is a logical subject with understandable reasons for everything. They should approach algebra with confidence that they have the ability to understand those reasons.

When exploration and reasoning do not lead to the expected results, students must be willing to go back and resolve the differences between their results and the expected results. They may have to start over again, or they may have to seek help. They should recognize that trial and error is a valid mathematical technique. They should not allow fear of errors to deter them from making attempts.

Some students may be studying hard but still not be learning; those students probably need to study smarter rather than harder. They need to be able to talk to someone about algebra. They should begin by doing their homework as early as possible and seeking help when they encounter something they do not understand. The first source of help is the teacher. Students should not be afraid to ask questions. In algebra, the "dumb questions" are the ones that should be asked but are not asked. Students should find other students with whom to study. Then when they get stuck on a problem, they have someone to talk to about it at school or someone to call from home.

Recently at the University of California, Uri Treisman conducted a study of factors that determined success of calculus students at the Berkeley Campus. He observed that race, sex, and socioeconomic status are not critical factors. The one critical factor was whether the students participated in study groups. He observed that whereas Asian students tended to work together in study groups, black students tended to study alone. He was able to get black students to work together through forming the Black Honors Calculus Society, and members' grades increased dramatically (Shanker 1988). A lesson can be learned here in expectations and communication that could apply to everyone who takes algebra.

Expectations and Responsibility for Learning Algebra

Beginning with the expectation that nearly all students can learn algebra puts more pressure on teachers. If students fail, teachers can't just shrug their shoulders and say that the students did not have the necessary ability because learning algebra is a shared responsibility between the teacher and the student. The teacher needs to assess and maintain prerequisite knowledge. The teacher needs to motivate students to study algebra through informing them of the importance of algebra to their future, through teaching them study skills, and through establishing an instructional setting in which students who apply themselves can be successful. Teachers need to use a variety of techniques to assess continually what students know, including the techniques of questioning and observation, as well as correction of homework and tests. Teachers must start from where students are to bring them to an understanding of algebraic concepts. Teachers can make algebra come alive by moving away from teaching for rote learning of mindless manipulation toward inspired teaching that conveys the power, the utility, and the beauty of algebra.

A teacher who has high expectations for students shares the responsibility for their success. Rather than ask why a student is unable to learn a concept, the teacher asks what changes in teaching strategy would enable the student to become successful. The teacher gives explanations and presentations that are adapted to the students' needs with a sensitivity to the levels of abstraction at which the students are functioning. The teacher makes liberal use of concrete materials and visuals to help the students understand the concepts being presented.

Expectations and Structuring Lessons for Success

A teacher with high expectations teaches lessons that are structured for success. A chapter or a lesson can be started by stating what is to be learned and why it is to be learned and by relating it to what came before and what will come later, with many of the lessons motivated by interesting applications or problems. The teacher ensures that the students have the necessary prerequisite knowledge for the lesson, and the direct-instruction portion of the lesson relates the new knowledge to the students' existing knowledge. Sample problems are demonstrated and discussed. The teacher checks for understanding. The teacher offers guided practice in class to ensure that students can successfully complete their assignments independently. Then the assignment should result in the students' having a successful experience applying the knowledge that was presented in the lesson. All too often, the assignment is made before the students have internalized the new material to be learned; then the teacher needs to explain the lesson while going over the previous day's homework. In the process, students begin with an unsuccessful experience that can contribute to lowered expectations by both the students and the teacher. Ideally, students should have success with every lesson; they should also be given some challenges with which they may not initially be successful. They need to know that the answers do not always come simply and easily; they should be encouraged to develop persistence and problem-solving skills to enable them to attack and ultimately to solve difficult problems.

The teacher should conduct frequent systematic reviews of previously learned material where individual students are held accountable for maintaining previous learning. A systematic review-and-maintenance program will help ensure that students always have the necessary prerequisite knowledge for future lessons.

Expectations and Assessment

A teacher with high expectations ensures that the tests measure objectives that have been taught and that the students have learned the material before being tested.

The teacher continually assesses students' progress through questioning, direct observation, and use of results of homework. The teacher modifies instruction as needed. Before taking a test, students should have a clear idea of exactly what is to be tested so both students and teachers are confident that the material has been learned. Surprises and trick questions should be avoided, but challenging questions that not all students are expected to answer correctly can be included and should be labeled as such. Before taking tests, students should be taught how to review for a test and how to take a test.

Expectations and the Learning Environment

The classroom of a teacher with high expectations is an orderly and friendly place in which students are expected to have consideration for each other; students should be expected to contribute to each other's learning. A teacher with high expectations makes every minute count for learning, with students kept on task from opening bell to closing bell (Johnson 1982).

A teacher with high expectations creates a nonthreatening learning environment in which students are encouraged to ask questions and take risks. Students should have opportunities to talk to each other about mathematics in small groups. They may be afraid to ask questions and appear stupid in front of a full class of their peers, but they are more willing to open up in small groups.

Cooperative learning is a practice that has great potential for enabling more students to learn algebra. In cooperative learning, students work together in groups of two to six students with a structure that rewards each member of a group when the entire group is successful. For example, the group can study together, take tests individually, and then each member of the group receives bonus points if all of the members of the group achieve above a certain score. This structure adds students' expectations for each other to individual student expectations and teacher expectations. Cooperative groups should be structured by the teacher to be heterogeneous by sex, ethnic background, and achievement level. Results from research indicate that cooperative learning promotes achievement, positive self-concept, and mutual understanding (Johnson & Johnson 1988–89). Cooperative learning is efficient because it uses the resource of the students' helping each other; when the teacher is called to give assistance more than one student at a time is helped. In cooperative groups students have opportunities to verbalize mathematics and to work on more challenging problems than they would be able to solve individually.

Teacher Expectations and Treatment of Students

Studies have revealed patterns of differential treatment by teachers of students for whom they have different expectations (Good 1987). For example, students with lower expectations tend to be seated farther away from the teacher, to be called on less often, and to have less attention paid to them by the teacher. The lower involvement of students with low teacher expectations is probably caused by both the student and the teacher. The student does not want to be bothered or embarrassed, and the teacher wants to focus efforts where they will do the most good.

In interacting with low-expectation students, the teacher gives less feedback, gives shorter waiting time for answers, and offers fewer clues to help the student with the answer. When the teacher has high expectations and believes that the student should be able to answer correctly, more wait-time is given and, if necessary, clues are provided. The teacher does not want to waste a lot of class time on a student who the teacher believes will not be able to answer correctly. The result is that the students who need the most help receive the least. Teachers can eliminate the differential wait-time problem by first asking the question, then allotting ample wait time for the whole class to think about it, and finally naming the student who is being asked to respond. This process keeps the entire class alert.

Low-expectation students tend to be criticized more often, praised less often, and interrupted more often. The teacher may be interacting with the student under the assumption that the student is more interested in disrupting the class than in learning. Low expectations contribute to a deterioration of the learning environment and are detrimental to learning.

Finally, teachers may demand less effort from low-expectation students and reward them for marginal responses. The teacher may be well-meaning, but the results can be devastating. Students may receive the message that the teacher doesn't expect as much from them as from other students and, therefore, that they don't need to put forth as much effort. The students may also receive the message that since acceptable work is not being required, the teacher has given up on them.

Many teachers do not treat students differently according to their expectations, but some do exhibit some of these characteristics. Teachers can identify and correct any differential treatment they have of students through a process of asking an observer to visit their classes and record their interactions with students for whom they have different expectations. Then the teacher can work on any problems that are identified.

Expectations, Algebra, and the Curriculum

To meet the expectation that most students can be successful in algebra and then go on to higher mathematics courses, some curriculum realignment must take place. The elementary school curriculum should move away from rote learning of computation to a focus on developing understanding of the mathematical concepts needed for learning algebra and geometry. A recent review of mathematics textbooks examined the percentage of pages at each level that contained any new material not contained in previous levels. The study revealed that in mathematics textbooks the percentage of pages of new material gradually decreases from a high of 100 percent in kindergarten (there is no previous level) to a low of about 30 percent in eighth grade. Then in algebra, which is usually taken in ninth grade, the percentage of new material jumps to about 90 percent (Flanders 1987). No wonder elementary algebra is so critical in the filtering process. A different focus in grades K–8 can help turn algebra from a filter to a pump.

Along with high expectations comes the need to give the algebra curriculum new focus to offer students instruction that is relevant to their future and eliminate content that is irrelevant and unnecessary. The algebra curriculum should emphasize understanding of algebraic concepts; applications of algebra in science, business, and other fields; and relationships between algebra and geometry. In the future, most symbol manipulation will be done electronically, so topics like factoring and simplification of rational expressions can be de-emphasized.

Algebra and Parental Expectations

Parental expectations can play a critical role in determining whether a student successfully completes algebra. Parents in upper-middle-class families tend to expect their boys and girls to complete successfully elementary algebra and higher level mathematics courses. They usually encourage the students to enroll in algebra and insist that they apply themselves. These parents are most likely to visit school and to be in touch with teachers, counselors, and administrators. If a student is not doing well, upper-middle-class parents are likely to provide direct help or arrange for tutoring.

Students whose parents have limited education may be at a disadvantage because of lack of parental support. However, that situation can be overcome. The school should let the parents know that algebra is extremely important to their children's future and that the students can learn algebra if they apply themselves. The most important educational role that a parent plays is in the encouragement of the students and the monitoring of their studying. Parents do not need to know algebra to help their students learn algebra. Comparative studies involving parents from the United States, Japan, and Taiwan present evidence that the monitoring role of parents is much more important than the tutoring role (Stevenson 1986).

Having high expectations for algebra students implies that a creditable standard of achievement will be expected of all students who take algebra. We must define and assess achievement on the basis of essential learning that students will need for the future. Students are hurt if they are given algebra credit when they have not learned the material that will be needed in later courses. They will have been dead-ended because they lack the empowerment for later success.

Multiple Opportunities for Learning Algebra

Although a common standard should be adopted for students receiving credit in algebra, we need to recognize that even though the same finish line should be used for all, not everyone is starting at the same place. Nor is it necessary for everyone to finish at the same time. Previously the first-year algebra course was placed at the ninth-grade level and students who did not complete elementary algebra by the end of ninth grade terminated their formal mathematics experience. The expectations and opportunities for these students to participate in higher learning were severely restricted. Minority students and students from low-income families tend to be disproportionately represented in this group of students. To remedy this situation we need to make special efforts to raise the expectations for groups of students who are underrepresented in algebra; we also need to take extra measures to ensure that they have the necessary prerequisite knowledge.

The opportunity to study algebra should always be available. One of the strengths of the educational system in the United States is the many opportunities it affords students who have not previously been successful. We must never give up on students or allow them to give up on themselves. With high expectations, encouragement, and effective instruction, the algebra course can be converted from a filter that screens people out to a pump that propels people forward toward opportunity.

REFERENCES

Armstrong, J., and S. Kahl. *An Overview of Factors Affecting Women's Participation in Mathematics.* Denver, Co.: National Assessment of Educational Progress, 1979.

Cheek, Helen N., Gilbert J. Cuevas, Judith E. Jacobs, Genevieve Knight, and B. Ross Taylor. *Handbook for Conducting Equity Activities in Mathematics Education.* Reston, Va.: National Council of Teachers of Mathematics, 1984.

Good, T. L. "Two Decades of Research on Teacher Expectations: Findings and Future Directions." *Journal of Teacher Education* 38 (1987): 32-47.

Flanders, J. R. "How Much of the Content in Mathematics Textbooks Is New?" *Arithmetic Teacher* 35 (September 1987): 18–23.

Johnson, David R. *Every Minute Counts: Making Your Math Class Work.* Palo Alto, Calif.: Dale Seymour Publications, 1982.

Johnson, David W., and Roger T. Johnson. "Using Cooperative Learning in Mathematics." In *Cooperative Learning in Mathematics,* edited by Neil Davidson, pp.103–25. Menlo Park, Calif.: Addison-Wesley, 1988.

______ ."Cooperative Learning in Mathematics Education." In *New Directions for Elementary School Mathematics,* 1989 Yearbook of the National Council of Teachers of Mathematics, edited by Paul R. Trafton, pp. 234-45. Reston, Va.: The Council, 1989.

National Council of Teachers of Mathematics. *Curriculum and Evaluation Standards for School Mathematics.* Reston, Va.: The Council, 1989.

National Council of Supervisors of Mathematics. "Essential Mathematics for the Twenty-first Century: The Position of the National Council of Supervisors of Mathematics." *Mathematics Teacher* 82 (September 1989): 470–74.

National Research Council. *Everybody Counts:A Report to the Nation on the Future of Mathematics Education.* Washington, D.C.: National Academy Press, 1990.

Sells, L.W. "High School Mathematics Enrollment by Race and Sex." In *The Mathematics Filter: A New Look at an Old Problem.* Hartford, Conn.: S. Hamer Associates, 1978.

Shanker, A. "Strength in Numbers." In *Academic Connections* [The College Board] Fall 1988, p. 12.

Stevenson, Harold W., Shin-Ying Lee, and James W. Stigler. "Mathematics Achievement of Chinese, Japanese, and American Children." *Science* 231 (1986): 693–99.

Taylor, B. Ross. "Equity in Mathematics: A Case Study." *Mathematics Teacher* 76 (January 1983): 12–17.

6

INSTRUCTIONAL STRATEGIES AND DELIVERY SYSTEMS

FRANKLIN D. DEMANA
BERT K. WAITS

CALCULATORS AND COMPUTERS make it possible to use realistic and interesting problems and applications as a means to learn mathematics. These technologies can help students understand mathematical processes, develop and explore mathematical concepts, and create algebraic and geometric representations of problem situations; they can turn the mathematics classroom into a laboratory in which students are active partners in the learning process. Numerical and graphical problem-solving techniques become accessible strategies for all students through the use of technology. Preliminary evidence indicates that technology helps make algebra accessible to all students and lays the foundation that makes the study of calculus and science successful. This approach holds special promise for traditionally underrepresented groups in mathematics and science.

DEVELOPING AND EXPLORING MATHEMATICAL CONCEPTS

Calculators can be used to help students understand the order of operations (Comstock and Demana 1987). Calculators with an Algebraic Operating System (AOS) should be used because many errors in algebra result from misunderstandings about the order of operations, and low-cost four-function chain-operation calculators tend to reinforce these misunderstandings. A chain-operation calculator is one that performs operations in the order in which they are received. For example, the keying sequence [2] [+] [3] [×] [4] [=] on a four- function calculator suggests that the value of 2 + 3 x 4 is 20. An AOS calculator gives 14, the correct value of 2 + 3 x 4.

Example 1. Write the display below each step in the following AOS calculator keying sequence.

Keying sequence: [2] [×] [3] [+] [4] [×] [5] [=]
Display: — — — — — — — —

(*a*) What does the calculator do when the [+] key is pressed?

(*b*) What does the calculator do when the second [×] key is pressed?

When the [+] key is pressed in the foregoing keying sequence, 6 will appear in the display. Thus, on an AOS calculator, the multiplication represented by the first [×] key is performed when the [+] key is pressed. When the second [×] key is pressed we see the previously entered 4. Thus, on an AOS calculator, the addition represented by the [+] key is *not* performed when the second [×] key is pressed. By observing the display on an AOS calculator it can be shown that multiplications and divisions are performed first, in order, from left to right and then additions and subtractions are performed. This sequence is the usual rule in mathematics for the order of operations. Students can use an AOS calculator to explore and establish the rules for the order of operations. We recommend extreme care when using the four function chain-operation calculator with young pupils because of the potential for creating misconception about the order of operations.

Students often give the incorrect value of 36 for the following algebra problem: Find the value of $2x^2$ when $x = 3$. Students make the mistake of first multiplying 2 by 3 and then squaring the resulting product, that is, they perform the operations in order from left to right. The next example shows how AOS calculators can be used to foreshadow the study of algebra by giving practice designed to guard against this type of error.

Example 2. Write the display below each step in the following AOS calculator keying sequence.

Keying sequence: [2] [×] [3] [x^2] [=]

Display: __ __ __ __ __

(*a*) What does the calculator do when the [x^2] key is pressed?
(*b*) What does the calculator do when the [=] key is pressed?

When the [x^2] key is pressed in the foregoing keying sequence, the displayed number is 9. Thus, students can see that only the 3 is squared. When the [=] key is pressed students can conclude that first the 3 is squared and then the result is multiplied by 2. Students trained this way are considerably less likely to make an order-of-operations error.

Calculator-and computer-based graphing can be used to help students develop significant understanding about graphing. Graphs of functions, relations, parametric and polar equations, and even three-dimensional graphing, are easily accessible with the aid of technology (Waits and Demana 1989). The next example illustrates how beginning algebra students might get started.

Example 3. Compare the graphs of the following functions.

(*a*) $y = x^2$, $y = x^2 + 4$, $y = x^2 - 3$
(*b*) $y = x^2$, $y = (x - 3)^2$, $y = (x + 4)^2$
(*c*) $y = x^2$, $y = 3x^2$, $y = 0.5x^2$
(*d*) $y = x^2$, $y = -3x^2$, $y = -0.5x^2$

The speed of calculator-or computer-based graphing makes possible the explorations suggested by example 3. In a matter of minutes students can obtain enough graphical evidence to make conjectures and then test them quickly with additional examples. As the graphs of the three functions in part (*a*) appear one at a time in the same viewing rectangle (fig. 6.1), students are able to discover the effect of adding a positive or negative number to the rule of a function. A viewing rectangle $[a, b]$ by $[c, d]$ is the rectangular portion of the coordinate plane determined by $a \leq x \leq b$, $c \leq y \leq d$ in which a graph is viewed. As the graphs of the three functions in (*b*) appear one at a time in the same viewing rectangle (fig. 6.2), students are able to discover that the effect of replacing x by $x + a$ in $y = x^2$ is to shift the graph of $y = x^2$ to the *left* a units if a is positive, and to the *right* $|a|$ units if a is negative.

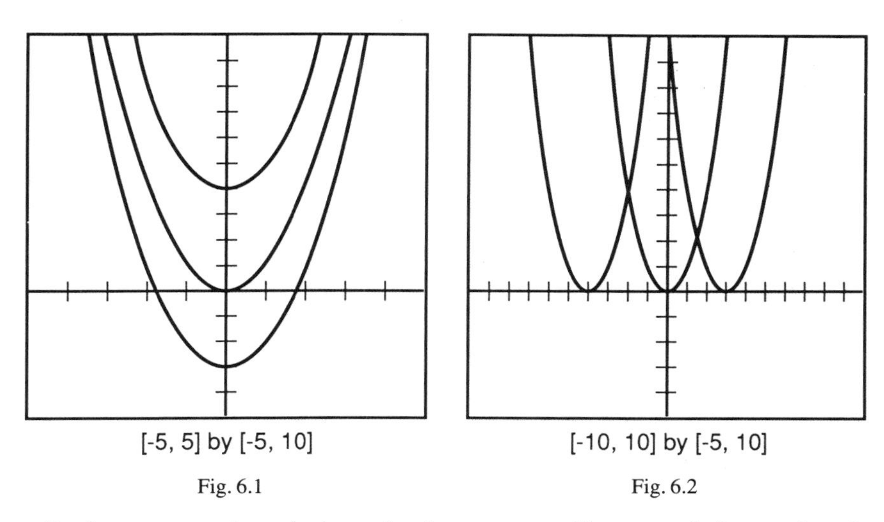

Fig. 6.1

Fig. 6.2

Students can continue the investigations suggested by example 3 to explore the horizontal and vertical shifting and stretching techniques required to explain how the graph of $y = a(x + b)^2 + c$ can be obtained from the graph of $y = x^2$. These investigations can be continued with polynomials of higher degree and other functions like $y = \sqrt{x}$, $y = |x|$, and $y = 1/x$. Transformation-graphing techniques will be well understood when the study of transcendental functions is started and, thus, the time required to study similar techniques with the trigonometric functions will be greatly reduced (Demana and Waits, 1990).

Instead of conventional lectures, teachers can design a careful sequence of problems to help students discover or explore important mathematical concepts. This type of activity turns the mathematics classroom into a laboratory in which students actively participate in the educational enterprise.

Students can use features called zoom-in and zoom-out to turn computers and graphing calculators (pocket computers) into efficient and powerful devices for student exploration (Demana and Waits 1988a). Zoom-in permits a portion of a graph to be "blown up" and analyzed. Zoom-in can be used as an extremely powerful and general method of solving equations and inequalities and finding maximum and minimum values of functions without calculus. Zoom-in is a fast geometric refinement of such numerical-approximation techniques as the bisection method. Zoom-out is a way quickly to determine the global behavior of a graph by viewing it in several large viewing rectangles, each containing the previous one. The behavior of a relation for $|x|$ large is called the *end behavior of the relation.* For example, students can use zoom-out to discover a general relationship for the end behavior of a rational function (Demana and Waits 1988b). The zoom-in and zoom-out features give a "complete" picture of the behavior of graphs of relations. The keys to successful implementation of this approach are the speed and power of a computer.

Scale

The use of technology in algebra focuses a great deal of attention on issues about scale. For example, the shape of graphs of specific functions like $y = x^2$ is a function of the viewing rectangle in which they are viewed. This phenomenon can be unsettling to teachers at first. The graph of $y = x^2$ in figure 6.1 is the one that teachers and students expect to see. The graphs in figures 6.3 and 6.4 are also of $y = x^2$, but they appear to be different because of the corresponding viewing-rectangle parameters.

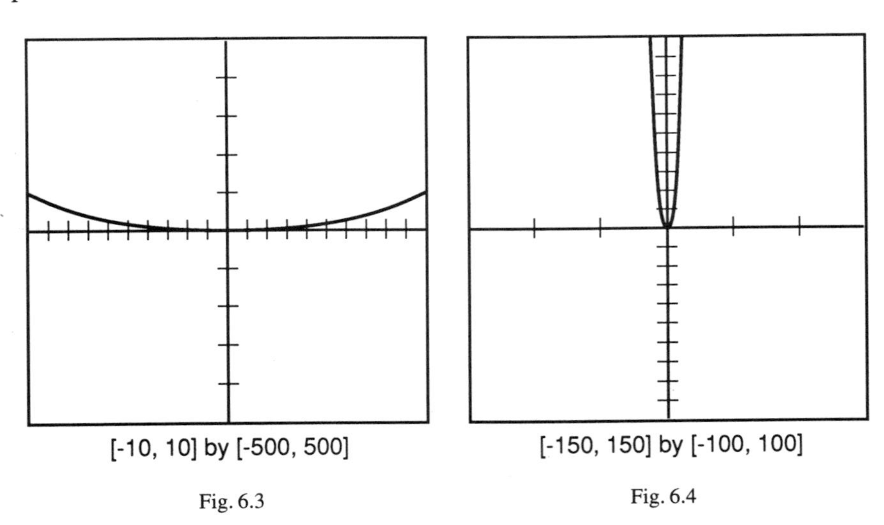

[-10, 10] by [-500, 500]

Fig. 6.3

[-150, 150] by [-100, 100]

Fig. 6.4

The next example illustrates that one must be careful that the visual representation one sees on a screen is actually true.

Example 4. Determine the behavior of

$$f(x) = \frac{2x^4 + 7x^3 + 7x^2 + 2x}{x^3 - x + 50}$$

in the interval [–2, 1]. The graph of the function *f* of example 4 in the [–20, 20] by [–20, 20] viewing rectangle is shown in figure 6.5. This graph leaves considerable doubt about the behavior of f between $x = -2$ and $x = 1$. If we zoom in a few times between $x = -2$ and $x = 1$ we get the graph shown in figure 6.6.

Notice that the function f has two local minima and one local maximum between $x = -2$ and $x = 1$. These local extremes are very close to each other, which causes this behavior to be hidden unless we observe the graph in a very small viewing rectangle. Although computer graphing is a very powerful and important tool, it does not remove the need for the study of algebra and advanced mathematics. Notice that the horizontal length of the viewing rectangle in figure 6.6 is forty times as large as the vertical length. It is often necessary to use nonsquare viewing rectangles to produce useful graphs.

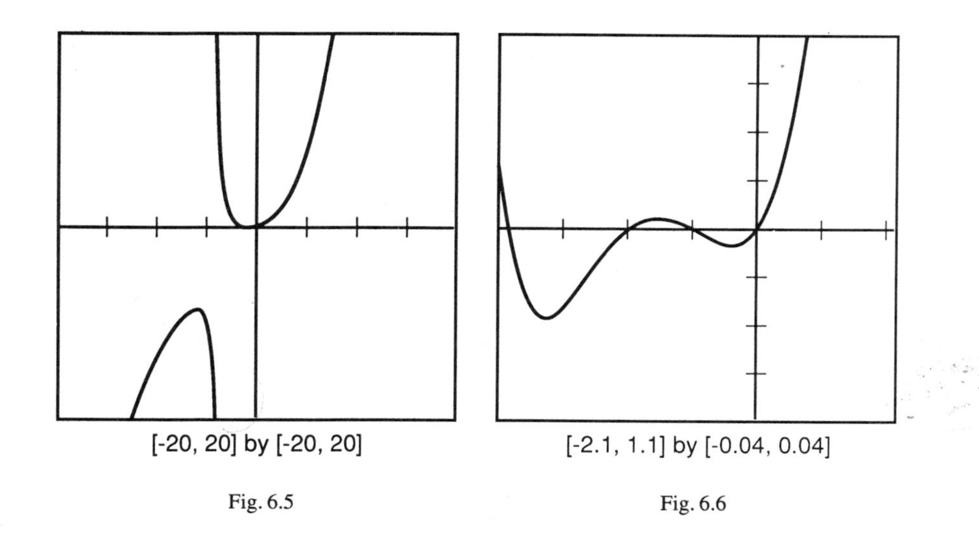

Fig. 6.5

Fig. 6.6

ALGEBRAIC REPRESENTATIONS FOR PROBLEMS

Calculator-based table building can be used to establish mathematical processes, to develop deeper understanding about variables, to solve problems numerically, and to help students write algebraic representations for problem situations (Demana and Leitzel 1988a). Interesting and realistic problems can be used to promote the study of mathematics because calculators remove the limitation of pencil-and-paper computation. Real-world problems are important to capture students' interest and to demonstrate the importance of mathematics. This activity will be illustrated with a problem about discount that has been used successfully with seventh- and eighth-grade prealgebra students.

Example 5. Starks Department Store is having a 32%-off sale.

(*a*) Complete the table on the top of the next page:
(*b*) Find the original price of an item if the discount is $16.
(*c*) Find the original price of an item whose sale price is $25.

Original Price ($)	Discount ($)	Sale Price ($)
10	$0.32(10) = 3.20$	$10 - 0.32(10) = 6.80$
20		
30		
40		
P		

Many students do not know how to find a given percent of a number. Teachers still need to focus attention on the process used to determine the entries in the second column of the table in part (*a*). However, freed from time-consuming pencil-and-paper computation, students can do numerous computations quickly with a calculator to grasp firmly that "thirty-two percent of a number" means to multiply the number by 0.32. Conventional paper-and-pencil techniques simply do not allow enough practice time for most students to fix this process in permanent memory. Notice how the last entry of the second column helps students see that one use of variable is as a generalization, or summary, of many numerical statements. This example gives students a deeper understanding about variable. The table itself furnishes concrete numerical representations of functions. The current curriculum pays almost no attention to the important and subtle concepts of variable and function. These gaps in the curriculum are responsible for blocking the mathematical progress of an incredible number of students.

The numerical entries in the second column are 3.20, 6.40, 9.60, and 12.80. Many students will discover the pattern of the entries of the second column of the table. Such students will see that the answer to part (*b*) is $50. For most prealgebra or beginning algebra students, solving part (*b*) is a guess-and-check activity. The completed entries in the second column can be used to make an initial estimate, and then guess-and-check used to solve part (*b*) by numerically zooming in on the answer.

The problem in part (*c*) is a typical word problem that algebra students often are not able to solve. More to the point, when algebra students are not able to solve this problem, they don't even know how to get started. Again, teachers can emphasize process in completing the entries of the third column of the table. This activity should help students see that the entry in the last row is $P - 0.32P$. Usually several students will offer $0.68P$ as an equivalent answer. Then, teachers can direct a lively classroom discussion of why these expressions are equivalent. Students use arithmetic evidence to convince themselves that these two expressions are equivalent. Teachers can use the distributive property to simplify both the numerical and algebraic entries in the last column. Here algebraic drill is buried in an interesting, realistic problem.

The last entry of the third column of the table above can be used to write an equation for part (*c*): either $P - 0.32P = 25$ or $0.68P = 25$. These equations are algebraic representations for the 32%-off sale problem. Calculator-based table-building activities help students produce algebraic representations for problems. Teachers will find that a great deal of algebraic "drill" can be disguised, and therefore practiced, in interesting calculator- or computer-generated activities—a frequent outcome of the use of technology in the mathematics classroom. Students seem to be receptive to algebraic activity when it is not the focus of a lesson.

GEOMETRIC REPRESENTATIONS FOR PROBLEMS

Prealgebra and beginning algebra students need to make some graphs by hand before using devices that automatically produce a graph of a function. For example, prealgebra students can produce a graph that shows how the sale prices of example 5 depend on the original prices using tables similar to the one in part (*a*). First they draw the points determined by their tables. Further analysis of the situation helps students see that the graph contains many more points than the ones given by their table. Finally, they should see that drawing the finished graph is equivalent to knowing the sale prices for all possible original prices. This use of concrete problems helps students see that the completed graph, a geometric representation of the 32%-off sale problem, is the portion of a straight line $y = 0.68x$ that lies in the first quadrant. Students then use their graphs to solve such problems as the one raised in part(*c*) of example 5.

Special attention must be paid to scale because students often draw their graphs on a small portion of the graph paper, or their graphs run off the graph paper. After some practice, students use the numbers in their tables to choose a scale (often different) for each axis that makes efficient use of their graph paper. This efficient use of graph paper can lead to the development of a considerable amount of number sense.

A good deal of effort is needed to help students establish the connections among the problematic situation, an algebraic representation of the situation, and a geometric representation of the situation—particularly with young students or students with little or no graphing experience. These connections will be discussed in more detail in the next example.

Once students understand how to make graphs, then calculator- and computer-based graphing can be used to produce graphs quickly enough to make graphing an effective problem-solving strategy. Producing graphs by hand takes so long that it is not possible to use graphing routinely to solve problems. With a graphing utility, any device that automatically produces graphs, graphing is a fast, effective problem-solving strategy, as the next example illustrates.

The following problem about making a box from a rectangular piece of cardboard can be started with prealgebra students and continued in algebra and precalculus classes to foreshadow the study of calculus.

Example 6. Squares of side length x are cut from each corner of a rectangular piece of cardboard that is 30 inches wide by 40 inches long (fig. 6.7). Then the cardboard is folded up along the dashed lines in figure 6.7 to form a box with no top.

(*a*) Write an equation (algebraic representation) that shows how the volume V of the box depends on x.

(*b*) Graph the equation in part (*a*) and indicate which portion of the graph represents the box problem.

(*c*) Determine the length of the side of the square that must be cut out to form a box with maximum possible volume, and determine the maximum volume.

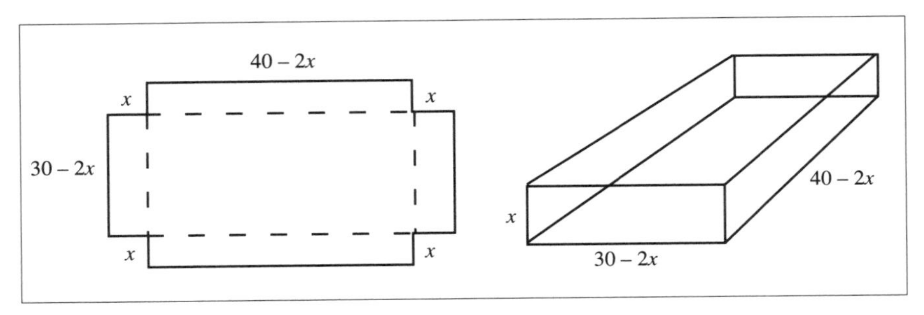

Fig. 6.7

Prealgebra students begin with writing an algebraic expression for the volume of the box using a table formed as in example 5. Finding the height, width, length, and volume for a few values of x leads students to see that the volume of the box is given by $V(x) = x(30 - 2x)(40 - 2x)$. Finding a viewing rectangle that shows the complete behavior of V is not an easy task. Notice that $V(1) = 1064$. Students don't expect values of a function to be that large for small values of the dependent variable. After some experimenting and teacher-guided investigation, students find that the graph of V in the [–10, 30] by [–3000, 5000] viewing rectangle (fig. 6.8) gives a rather complete picture of the behavior of V. This graph of the algebraic representation $V(x) = x(30 - 2x)(40 - 2x)$ of the box problem is one possible geometric representation of the problem. Before using graphs to answer questions about the box problem, teachers must help students establish the necessary connections.

Deciding which part of this graph represents the problem can generate a lot of quality student-teacher discussion. First, students must be guided to see that if (a, b) is a point on the graph of V, then a is a possible side length of the square that is cut from the corners of the rectangular piece of cardboard and b is the corresponding volume of the box that is formed. This realization allows students to eliminate portions of the graph of V as not representing the situation in the problem. The portion of the graph beyond the largest x-intercept may cause the most trouble. Eventually students understand that the graph in figure 6.8 represents the box problem only for x between 0 and 15 (fig. 6.9).

Next the zoom-in feature can be used to find the coordinates of the highest point of V in figure 6.9. The x-coordinate of this point is the side length of the square that must be cut out to produce a box with maximum volume, and the y-coordinate of this point is the corresponding maximum volume of the box. This result gives the answer to (*c*) of example 6. The coordinates of this point are approximately (5.66, 3032.3). Thus, if a square of side length 5.66 inches is cut out, a box with volume 3032.3 in^3 is produced, which is the largest possible volume for such a box.

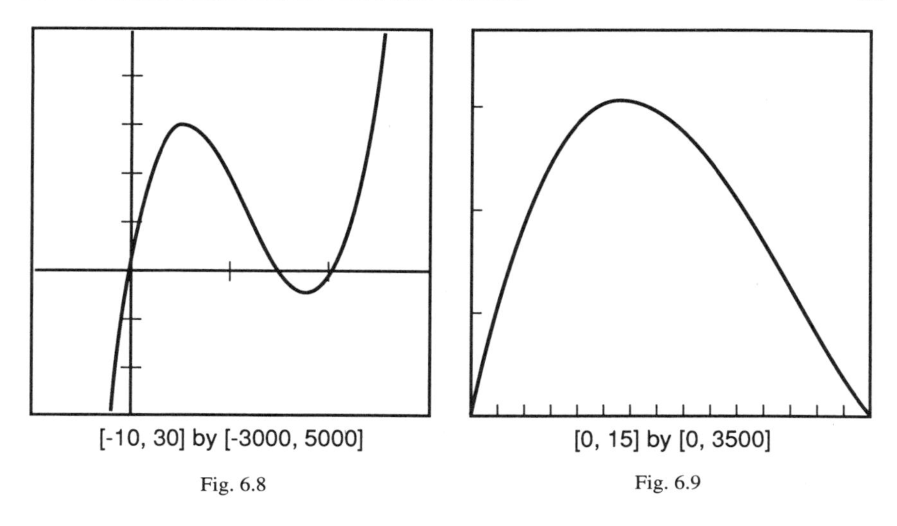

Fig. 6.8

Fig. 6.9

SUMMARY

The use of technology in instruction has many important consequences. Students view the study of mathematics as important because they are able to solve realistic, real-world problems. The lack of arithmetic and algebraic facility is no longer a barrier to mathematical progress. The focus of algebra shifts from computation and manipulation to the use of algebra as a language of representation. Technology has the important by-product of helping students improve their arithmetic and algebraic skill by giving geometric and realistic meaning to these usually mindless tasks. Many more students will now be able to pursue careers in science- and mathematics-related fields because technology makes mathematics accessible to many more students—particularly students that have traditionally excluded these options because of the roadblock caused by mathematics.

REFERENCES

Comstock, Margaret, and Franklin Demana. "The Calculator Is a Problem-solving Concept Developer." *Arithmetic Teacher* 34 (February 1987):48–51.

Demana, Franklin D., and Joan R. Leitzel. "Establishing Fundamental Concepts through Numerical Problem Solving." In *The Ideas of Algebra, K–12*. 1988 Yearbook of the National Council of Teachers of Mathematics, 61–68. Reston, Va.: The Council, 1988a.

———. *Getting Ready for Algebra—Level 1 and 2*. Lexington, Mass.: D. C. Heath & Co. 1988b.

Demana, Franklin D., and Bert K. Waits. "Microcomputer Graphing—a Microscope for the Mathematics Student." *School Science and Mathematics* 88 (March 1988a): 218–224.

———. "The Ohio State University Calculator and Computer Precalculus Project: The Mathematics of Tomorrow Today!" *AMATYC Review* 10 (Fall 1988b): 46–55.

———. *Precalculus Mathematics, a Graphing Approach*. Reading, Mass.: Addison-Wesley Publishing Co., 1990.

Waits, Bert K., and Franklin D. Demana. Master Grapher. Computer Software for IBM, Apple II, and Macintosh. Reading, Mass.: Addison-Wesley Publishing Co., 1989.

7

COMMUNICATING THE IMPORTANCE OF ALGEBRA TO STUDENTS

Paul T. Christmas
James T. Fey

Every branch of mathematics furnishes ideas and techniques for representation of, and reasoning about, structural properties or patterns in observed or imaginary situations. The concepts, principles, and methods of algebra constitute powerful intellectual tools for representing quantitative information and then reasoning about that information. The central concepts of algebra include variables, functions, relations, equations and inequalities, and graphs. The central principles are the structural properties of the real-number system and its important subsets. Those concepts and principles combine to give a system of symbols for describing and drawing inferences from relationships among quantitative variables.

Students must represent and handle quantitative information throughout their elementary school study of arithmetic. They answer thousands of practical and whimsical questions that require ordering of, or operations on, numbers. But the generalization of that experience and skill to problems that require reasoning about "all numbers x" or "some (unknown) number x" is usually difficult for beginning algebra students. Those students who are mystified by the new abstract setting for thought frequently seek familiar arithmetic methods to answer the questions posed or resist instruction with the plaintive, "When am I ever going to need this?"

Traditional justifications for the study of algebra include promises that it is essential for learning advanced mathematics or demonstrations of word problems that ostensibly require algebra. The "delayed gratification" explanation works for a few dedicated students who are convinced that they will or must study mathematics for several more years. The typical algebra word problems concerning ages, digits, mixtures, and speeding airplanes might convince a few more (if they don't realize that in most problems some arithmetic trial-and-error is probably as effective as formal algebra). But vast numbers of less willing or less serious students remain

Research for this paper was supported in part by the National Science Foundation under NSF award no. MDR 87-51500. Any opinions or findings, conclusions, or recommendations expressed herein are those of the authors and do not necessarily reflect the views of the National Science Foundation.

unconvinced that the current algebra syllabus prepares them for any probable or important future.

What is important about elementary algebra as a subject to be learned by average and low-ability students? How can this importance be communicated to those students? It seems to us that elementary algebra has two fundamental aspects, and for each we have many different ways of illustrating the power and significance of learning the subject.

REPRESENTATION OF QUANTITATIVE PATTERNS

In every list of strategies for effective problem solving, the recommended first step is to understand the problem—the known and unknown information and the relations among factors in the problematic situation. To apply quantitative reasoning to such situations, it is almost always helpful to represent that factual and relational information in the compact, unambiguous language of mathematics—to create a *mathematical model* of the problem's conditions.

When the situation to be modeled includes unknown or variable quantities, effective mathematical representation requires the use of such fundamental algebraic concepts as variable, function, equation or inequality, relation, and graph and the symbolic notation for labeling those concepts. Thus algebra embodies many of the most important ideas and symbolic conventions in the language of mathematics. The ability to use both the concepts and symbolic conventions of algebra in this representation process is very useful in a broad range of situations that students of even modest mathematical ability and interest can find meaningful and interesting.

Formulas

The most common and familiar uses of algebra to represent mathematical relations are the many simple formulas that occur throughout business, industry, science, technology, and the personal decision making of daily life. For instance, the basic formula relating distance, rate, and time is used in an incredible array of situations, including the familiar textbook problems with speeding trains, planes, and automobiles and some other surprising but common phenomena.

Example

Sound travels at a speed of roughly 0.21 miles per second, but light travels at over 180 000 miles per second. The relation between distance and time for sound is $d = 0.21t$. The relation for light (and similar waves) is $d = 180\ 000t$. This difference explains some very interesting physical events.

The familiar calculation of distance for an approaching thunderstorm assumes that lightning, traveling at the speed of light, arrives almost instantaneously, whereas the accompanying thunder, traveling at the speed of sound, takes longer. Thus if thunder arrives 5 seconds after lightning is seen, the lightning must have struck approximately 0.21×5, or 1.05, miles away.

This same relation explains the surprising fact that the sound of a starter's gun in a race can be heard by radio listeners hundreds of miles away before it is heard by spectators on site who are only a fraction of a mile away. The radio wave travels at the speed of light!

Of course, the foregoing is only one example of the role that formulas play in representing important relations among quantitative variables in everyday situations. The formulas for area, volume, perimeter, interest, and many other routine calculations are strong testimony to the usefulness of algebraic notation and concepts.

Management Information Systems

Formulas for area, bank interest, temperature conversion, distance-rate-time, and other familiar quantitative relations have been offered as applications of simple algebra—and as arguments to convince reluctant learners—for a long time. Newer and less appreciated is the way that the more general ability to build and use formulas plays a fundamental role in the range of management-information systems used by businesses throughout our contemporary economy.

Example

The recent break-up of the Bell telephone system has led to a proliferation of service and pricing options, and those options change frequently. The information is collected and processed by computer models consisting of many related formulas like the following:

$$\text{monthly charge} = \text{base cost} + \text{evening rate} * \text{evening use} + \text{daytime rate} * \text{daytime use}$$

If you are one of the many telephone company employees, you would certainly find it helpful to understand how those formulas function and how they are represented symbolically in the system. If you are a customer, you would also find it helpful in dealing with the system to have some general understanding of the formulas and the ability to compare effects of different systems.

Variables, Functions, and Relations

The symbolic language of an algebraic model permits hypothetical, or "what if," reasoning about variables and relations in a situation. For instance, if a chain of stores that rent VCR tapes has formulas relating its prices, rentals, revenue, costs, and profits, managers can experiment mathematically to predict the effects of various decisions. In the increasingly quantitative world in which we live, an understanding of how such models can be used (and abused) has become practical knowledge for many business employees and a useful source of insight for intelligent consumers and citizens.

Example

One of the basic concepts of marketing is the demand equation—one that relates the price of a product and likely sales. The demand equation leads naturally to a revenue function and the opportunity to experiment with the effect of price on revenue.

If daily VCR rentals for a store are related to rental charge by the equation $y = 1250 - 500x$, then revenue is a function of price with equation $R = 1250x - 500x^2$. Inspection of the linear rule for demand shows that daily rentals decrease at a rate of 500 per $1 increase in price. Numerical experimentation will show that decreasing pattern and also the effects of increasing price on revenue.

Although some might argue that this kind of economic analysis is important only for a few specialists in each business, the basic principles seem useful for many more employees and customers; and any very clear understanding of those principles really depends on the ability to represent the relation algebraically.

An understanding of the concepts of variable and function, and the cause-and-effect relations they represent, can be developed very effectively by use of the intriguing computer simulations that are now available. For instance, the original Apple program Lemonade Stand gave students a captivating setting for "What if?" experimentation. That classic software has now been joined on the market by many other simulations, some of which can be analyzed by studying the functional relations used in converting input decisions to outcomes.

Simulation is a very common tool of business, industrial, and scientific planning today. In every situation, to create a suitable mathematical model that simulates the system's behavior, it is essential to represent quantities or objects (e.g., cars in a computer wind-tunnel or homes on the route of a trash-collection service) by numerical and algebraic expressions. It seems important to help many of our students to understand how those simulations work, at least in some general way, and also to appreciate the limitations of such abstract models.

Spreadsheet Models

The "what if" reasoning that depends on use of variables and functions from algebra is also commonly done with computer-spreadsheet models. The design and use of those spreadsheet models absolutely require an understanding of algebraic representation using symbolic expressions.

Example

The following two spreadsheets show the setup required to experiment with different monthly payments, interest rates, and length of mortgage. Entries in boldface are input values that the user can manipulate to see instant recalculation by the formulas.

(1) Showing cell formulas:

A	B	C	D
1 Monthly payment	**850**		
2 Interest rate	**11.5%**	Period rate	B2/12
3 Years of mortgage	**30**	Months	B3 * 12
4 Possible mortgage		B1 * (1 + (1 + D2) ∧(–D3))/D2	

(2) Showing cell values:

	A	B	C	D
1	Monthly payment	**850**		
2	Interest rate	**11.5%**	Period rate	0.009 583 3
3	Years of mortgage	**30**	Months	360
4	Possible mortgage		91 557.99	

This example shows how the ability to handle algebraic concepts and notation is essential in the use of spreadsheet models and, further, how those models can be used to illustrate the central concepts of algebra.

Finding Algebraic Models

For many algebra students the most frustrating part of problem solving is translating given conditions into appropriate symbolic function, equation, or inequality models. That frustration often feeds skepticism about the importance of the task when the problems are the standard puzzles involving ages, digits, mixtures, or work rates. In many genuine applications of algebra, the sources of function rules, equations, or inequalities are not carefully worded verbal statements but experiments in which data are collected and displayed before an algebraic model relating variables is sought. The use of such data-collection activities can present interesting and convincing opportunities to show the power of algebraic representation.

Example

Students can simulate the experience of a market-research study by doing a (price, sales) survey on some consumer item of interest to them and their classmates. For instance, in a class of twenty-five ask how many students would buy a popular tape or CD at various prices ranging from $1 to $15. Plot the (price, sales) data, and draw a line of reasonable fit. Then find the equation of that line.

Students can perform classroom experiments with physical apparatus to find relations like that between the height at which a ball is dropped and its rebound height. The simple strategy is to collect a number of (release, rebound) height pairs, plot those data pairs, and find an equation that fits the relation.

In each situation the rationale for finding an algebraic expression for the relation between variables includes efficiency (one equation is much more compact than a table of data) and insight (the coefficients in a linear equation $y = mx + b$ usually represent significant properties of the relation).

Multiple Representations

The preceding examples of ways in which experiments lead to data and then to algebraic models also illustrate one of the most powerful features of contemporary approaches to algebra—the use of calculators and computers to construct multiple representations of relations among variables.

Example

Suppose the owner of a chain of car-wash businesses has studied her business prospects and found that the average daily profit depends on the number of cars washed, with the modeling rule

$$P(n) = -0.027n^2 + 8n - 280.$$

To find the number of cars that will lead to break-even or maximum profit situations, it is helpful to study tables of $(n, P(n))$ values, a graph of $P(n)$, and the pattern in the rule itself. The ability to construct and analyze such multiple representations is a valuable skill for anyone who will face quantitative reasoning tasks.

Historical Perspective

Although many students are motivated to study a school subject by its immediacy to their personal lives, some are also intrigued by historical or futurist perspectives. In the history of science and mathematics, the evolution of algebraic notation and methods is one of the most interesting and important themes. Students who are reluctant to adopt the compact symbolic forms of contemporary algebra might have quite a different appreciation of their power after following a short trip from the literal style of Babylonian mathematics to the syncopated and geometric styles of Greek mathematics and on to the precursors of current usage in the mathematical work of the Renaissance.

Example

Here are only two of many stages in the evolution of modern symbolic notation:

cubus p̄ 6 rebus aequalis 20 (for $x^3 + 6x = 20$)

$aaa - 3\,bb ==== +2\,ccc$ (for $a^3 - 3b^2 = 2c^3$)

Even earlier work gives examples in which such equations were written in complete prose sentences.

Even a short exploration of algebraic representation from a historical perspective should help some students gain valuable insight into, and appreciation for, the meaning and power of modern abstract styles and the importance of representing ideas clearly and efficiently.

PROCEDURAL REASONING IN ALGEBRA

The use of algebraic notation to model quantitative relations is a powerful first step toward effective quantitative problem solving and decision making. However, once a question has been translated into an equation or inequality or a system of such conditions, the important and often difficult task of solving the algebra problem remains. Traditional algebra courses have always included, in fact emphasized, a vast array of routine procedures for manipulating symbolic expressions into different equivalent forms in the search for a solution. Students have been expected to become facile with operations on polynomial and rational expressions to perform the required simplifications or expansions.

Because this aspect of algebra is routine and rule-bound, it has been an intriguing challenge for computer software authors seeking to supply helpful tools for algebraic problem solving. Several very powerful computer programs do symbolic algebraic manipulations in response to such standard commands as SOLVE, FACTOR, EXPAND, and so on. These programs have many of the same implications for elementary algebra as hand-held calculators for arithmetic—diminishing the importance of procedural skill and highlighting the importance of problem formulation, estimation, and interpretation of results. Thus the traditional procedural part of algebra is, especially for students of modest ability, far less significant than the ability to construct and interpret algebraic representations of quantitative relations. Nonetheless, the importance of learning algebraic procedural skills can be argued in several ways.

Number Systems and Properties

In standard elementary school mathematics curricula, students meet and master properties of the whole-number and positive-rational-number systems. Not until the study of algebraic ideas are the properties and operations of negative and irrational numbers thoroughly developed. Although much of the factual and relational information in a quantitative problem can usually be modeled by use of positive numbers only, it is becoming increasingly common in computer-based systems to use negative numbers to represent inputs and outputs for model calculations. Thus the ability to interpret negative numbers representing business losses, time prior to a space-shuttle launch, position below sea level, and many other "opposite" quantities is useful. When these situations are being modeled by a computer system, it is not sufficient for a user to assume that the machine will "know what I mean." Data must be entered in a manner consistent with the modeling assumptions, and operations must be specified in the order that will produce intended results.

The conventional motivation for studying irrational numbers is the fact that the Pythagorean theorem leads to calculations involving square roots. For average and low-ability students, it is not at all clear that this fact of mathematical life justifies a full-blown treatment of radicals, fractional exponents, and so on. However, it seems, at a minimum, worth demonstrating numerically and then mentioning the fact that numbers like the square root of 2 and pi can only be approximated by decimals or common fractions like 1.414 and 22/7 and that the degree of accuracy required is a function of the demands of the particular situation. For instance, if one is designing a support wire for a 100-meter radio antenna, it will be sufficient to use approximations like 1.73 for the square root of 3. But in planning space travel to the moon, one could land far off the mark when using 22/7 for pi .

The key objective in teaching about negative and irrational numbers should be developing students' ability to set up and interpret mathematical models where those numbers are needed—not facile operations of arithmetic with these quantities. However, the analysis of number-system properties that is also a normal part of algebra has some significant payoffs in arithmetic calculation. In the quantitative reasoning tasks for which students are being prepared by school mathematics, most arithmetic computations will be done with the help of a calculator or computer. Despite this condition, nearly everyone in mathematics education has urged increased emphasis on the mental computation needed in estimation and approximate calculations that guard against errors of data entry or machine computation. The structural properties of number systems are enormously helpful in this approximate calculation.

Example

Properties of signed numbers permit rearrangements for easier calculations, as shown:

$$\begin{aligned} 50 - 23 - 34 + 75 - 18 &= (50 + 75) - (23 + 34 + 18) \\ &= 125 - 75 \\ &= 50 \end{aligned}$$

Example

For students who have learned arithmetic algorithms in a very rote fashion, revealing power is found in simple applications of the distributive property, as in the following derivation of a formula for compound interest on $500 invested at 8% annual interest:

$$\begin{aligned} 500 + (.08 \times 500) &= (1 \times 500) + (.08 \times 500) \\ &= (1 + .08) \times 500 \\ &= 1.08 \times 500 \\ &= 540 \end{aligned}$$

In exactly the same way, students can see that

$$\begin{aligned} 540 + (.08 \times 540) &= (1 \times 540) + (.08 \times 540) \\ &= (1.08) \times 540 \\ &= (1.08) \times (1.08) \times 500 \; [540 = 1.08 \times 500] \\ &= (1.08)^2 \times 500, \end{aligned}$$

and so on.

Example

Number-system properties and signed-number operations are the bases of other computational shortcuts, like the following quick estimate for an arithmetic mean:

The mean of 45, 52, 60, 48, and 68 appears to be about 55. To test this estimate, calculate as follows:

$$\begin{array}{ll} 45 - 55 = -10 & \\ 52 - 55 = -3 & (-13) \\ 60 - 55 = 5 & (-8) \\ 48 - 55 = -7 & (-15) \\ 68 - 55 = 13 & (-2) \end{array}$$

So the true mean is $55 + (-2/5)$.

Of course, if these and other computational shortcuts are used carelessly or remembered poorly, they are invitations to fatal errors. This fact argues for an approach that bases each proposed shortcut on the clear understanding of number-system properties that is part of algebra.

Manipulations Done Easier by Hand

As a practical matter, people who use algebra for quantitative modeling and problem solving in the future will undoubtedly rely on computer assistance for much of the symbolic manipulation that has been the heart of the traditional course. However, some situations are certainly so simple in structure that doing the manipulation by hand (perhaps with calculator assistance for the related arithmetic) is far more efficient than looking for a suitable machine. Furthermore, in those simple situations, developing the procedural skills required to solve equations or inequalities can be done in a way that strengthens understanding of the relational form and its use as a model.

Among the family of algebraic forms that students are expected to master in a traditional course, those that meet the "easier by hand" criterion certainly include—

1. linear equations and inequalities of the form

$$ax + b = c \text{ and } ax + b < c;$$

2. quadratic equations and inequalities of the form

$$ax^2 + b = c \text{ and } ax^2 + b < c;$$

3. rational equations of the form

$$a/x = b \text{ and } a/x^2 = b.$$

Linear equations and inequalities of the type described here include a vast majority of practically occurring situations in which linear relations are appropriate models of quantitative relations. Furthermore, the procedure for solving such linear equations can be naturally related to the operations on quantities being modeled by the function $f(x) = ax + b$.

Example

The cost of membership in a video club includes an annual fee of \$15.00 plus \$2.50 for each cassette rented for one day. Thus the annual cost is given by the function $C(n) = 15 + 2.50n$, where n is the number of cassette rental days used during a year. To answer a question like "How many cassette rental days can be used to keep annual cost under \$200?" one must solve the inequality

$$15 + 2.50n < 200.00.$$

To solve this problem one must reverse the sequence of operations needed to calculate cost from number of rental days, that is, find (200 – 15)/2.50.

The solving procedure follows naturally from the procedure for calculating output values from input values of n.

The quadratic cases listed in the foregoing provide some important practical models, too.

Example

If a ball is dropped from a tower that is 100 meters tall, its height after t seconds is given by the function $h(t) = -4.9t^2 + 100$.

To answer a question like "When will the ball hit the ground?" it is necessary to solve the equation $-4.9t^2 + 100 = 0$. In working backward from desired output to required input t, we find that the steps are identical to the linear case until we arrive at $t^2 = 20.408$.

Solving this equation requires an understanding of the squaring operation and one push on the calculator's square-root button. Again, the solution process reinforces the understanding of the procedure for calculating outputs from inputs for this function rule.

Although this form of quadratic does not cover all important quadratic relations, it has the virtue of building on the linear case and revealing the multiple-root behavior of quadratics. Furthermore, the methods for solving full quadratic equations hardly meet the criterion of "easier by hand," particularly for less able mathematics students.

Example

When sound emanates from the speakers of a rock band, its intensity diminishes with distance according to a function rule of the form $I(d) = a/d^2$, where intensity is in watts per meter squared and distance is in meters.

To answer a question like "If a sound measures 0.2 watts per meter squared one meter from its source, at what distance will it be reduced to 0.004 watts per meter squared?" one must solve the equation $0.2/d^2 = 0.004$. Again, reversing the "input to output" procedures reveals the answer, $d = \sqrt{50}$.

This example and many others like it cover the very important family of situations modeled by inverse variation. In addition to this sound-intensity setting, inverse variation occurs frequently in natural phenomena like light intensity, gravitational attraction, and "time as a function of rate" problems where some distance is to be traveled or a job is to be completed. As with the previous linear and quadratic examples, the procedures required are simple.

For any students who are studying algebra, even those of limited mathematical ability, it seems reasonable to argue that the few basic symbolic forms identified here are among those for which procedures that can be executed "by hand" are important to learn. Because they occur very often, they are, in fact, generally easier to solve by hand than by searching for a computer program with a SOLVE feature, and learning the natural solution procedures illuminates the structure of the relations being modeled.

The reader will notice that we have not suggested the typical symbol-manipulation procedures based on meticulous application of number-system features like the associative, commutative, distributive, inverse, and identity properties. For almost all algebra students this sort of formal approach to equation solving, although generalizable to cases of considerably greater complexity, quickly becomes just that—a formal exercise that seems to have little to do with the practical business of using algebra to model situations and solve meaningful problems. It does not seem important in presentation of elementary algebraic concepts and methods to students of modest ability and interest.

Algorithmic Methods

The study of school geometry is often supported by arguments that it develops students' logical reasoning ability. Algebra is seldom given the same kind of endorsement, but we have some reasons to believe that certain general habits of thought required in algebra might carry over to a broader range of intellectual tasks. For example, algebraic notation is among the most abstract, efficient, and powerful systems for expressing information. However, it also demands absolute precision in its use. It has none of the redundancy built into ordinary language, so it encourages care in expression and manipulation of ideas. This habit of carefulness is especially useful in the broad array of situations in which mechanical or computer systems are used as tools in some specific job or career. Although many such systems are now designed to help human users avoid dramatic errors, a miscue as simple as pressing the wrong key on an automated supermarket check-out system can produce troublesome difficulties.

In addition to the requirements for precision of expression, the variety of problem-solving procedures that constitute much of elementary algebra are illustrative of a general trait that characterizes methods of automated systems. The systems that dispense tickets to subway passengers, check out books for library users, send bills and statements to credit-card and bank customers, and control the flow of parts in a manufacturing process all follow precisely defined rules of operation called algorithms. In the daily life of contemporary society, many jobs require the ability to design those algorithms. Many more require working with good judgment alongside algorithm-driven automated systems, and nearly everyone encounters such systems as a consumer. When systems function as expected, they are not really even noticed; but when something goes wrong, some general understanding of how systems are run by algorithmic procedures will make the detection of flaws and their correction much easier tasks.

If algebraic procedures are taught with the proper attention to their place in the broader family of algorithmic methods—emphasizing the usually critical importance of order and precision and the fatal effects of even small errors—it seems quite possible that students of even modest mathematical ability can gain valuable insights into the way many systems that they will use and depend on are designed and function. Thus studying procedural aspects of algebra with even modest levels of complexity offers some impressive opportunities for development of important general thinking habits and skills.

Historical Perspective

As with algebraic representation, the history of efforts to develop algebraic procedures for quantitative problem solving contains a number of interesting and impressive themes. Showing students something of the outline of this story should help to illuminate the basic goals and fundamental difficulties in procedural thinking in algebra.

For instance, in early Babylonian mathematics—without the benefit of symbolic notation or signed numbers—solution procedures to verbally stated equations had to be described in prose sentences. During the golden age of Greek mathematics, solving an equation meant devising a straightedge-and-compass geometric construction of the required magnitude. In the Middle Ages, mathematicians competed with each other in solving equations that are now routine tasks; each specific equation with its particular coefficients was seen as a new problem because the general reasoning methods that we take for granted had not been abstracted and validated.

Although this long and difficult path to the powerful contemporary methods of algebra is not itself important for students to know, the telling of that story might, for some students, help convey the significance of the intellectual achievement that modern algebra represents.

SUMMARY

Algebra is clearly the backbone of secondary school mathematics. It furnishes concepts and symbolic conventions for representation of very important information in situations that affect each of us in obvious and subtle ways every day. Understanding of some basic ideas underlying that style of representing or modeling quantitative information is now a critical prerequisite for entry into many careers and for effective life in dealing with the quantitative-information systems that impinge on everyday affairs.

The procedural methods of algebra—the rules for transforming symbolic representations into equivalent but simpler patterns—are also widely used in the pervasive rule-driven systems that we see all around us. Although computerization makes many traditional, by-hand methods of symbolic manipulation less important for all (and certainly for less quantitatively able students), some important general lessons about precision of expression and algorithmic thinking can emerge from experience with learning algebraic methods.

Many sources are available from which teachers can draw examples illustrating the usefulness of elementary algebra. We list in the Bibliography only a few of the books from which ideas were drawn for this paper.

BIBLIOGRAPHY

Coxford, Arthur F., and Albert P. Shulte, eds. *The Ideas of Algebra, K–12*. 1988 Yearbook of the National Council of Teachers of Mathematics. Reston, Va.: The Council, 1988.

Foerster, P. *Algebra I*. Menlo Park, Calif.: Addison-Wesley Publishing Co., 1984.

Jacobs, Harold R. *Elementary Algebra*. San Francisco: W.H. Freeman & Co., 1979.

Kastner, Bernice. *Space Mathematics: A Resource for Secondary School Teachers*. Washington, D. C.: National Aeronautics and Space Administration, 1985.

McConnell, J., et al. *Algebra*. Chicago: University of Chicago School Mathematics Project, 1987.

Mathematical Association of America and National Council of Teachers of Mathematics (NCTM), Joint Committee. *A Sourcebook of Applications of School Mathematics*. Reston, Va.: NCTM, 1980.

National Council of Teachers of Mathematics. *Historical Topics for the Mathematics Classroom*. Thirty-first Yearbook. Washington, D.C.: The Council, 1969.

Usiskin, Zalman. "What Should Not Be in the Algebra and Geometry Curricula of Average College-bound Students?" *Mathematics Teacher* 73 (1980): 413–24.

8

LIST OF RESOURCES

DOROTHY S. STRONG

THE LIST OF RESOURCES takes into consideration the unique nature of the program *Algebra for Everyone*. Resources needed to support such an algebra program must get to the heart of the problem. Although materials for mathematics content and methodology are necessary, they are not sufficient to achieve the goal identified. Effective mathematics for everyone requires that teachers be familiar with the necessary mathematics content, appropriate methodology for teaching this content, the characteristics of the students to be taught, obstacles to success, and strategies for overcoming these obstacles. The list of resources identified for this program takes all these factors into consideration. This list is not exhaustive but includes examples of materials that are available in each category. The categories for which resources have been identified are expectations, equity, learning styles, cooperative learning, practices and programs, funding sources, instructional materials, manipulatives, and computer software.

Summarizing descriptions are given for each resource. Some resources are listed more than once, since they appear to belong to several categories. Descriptions are not included for second or subsequent listings. The instructional materials include manipulatives that are marketed by several companies; the company listed may or may not be the manufacturer of the item.

A summary chart, which follows immediately, identifies the materials and the categories in which they are listed. This chart is followed by an annotated list of the resources identified in the chart. It is hoped that teachers seeking materials to enrich their mathematics programs will find this list of resources useful.

Summary Chart

Title	Expectations	Equity	Learning Styles	Cooperative Learning	Practices & Programs	Funding Sources	Instructional Materials	Manipulatives	Computer Software
Mathematics and Science: Critical Filters for the Future	X								
"Rumors of Inferiority: The Hidden Obstacles to Black Success"	X	X							
Countering the Conspiracy to Destroy Black Boys	X								
Countering the Conspiracy to Destroy Black Boys: Volume II	X								
Motivating and Preparing Black Youth to Work	X								
Developing Positive Self-Images and Discipline in Black Children	X								
Women and the Mathematical Mystique	X								
The Mathematics Report Card: Are We Measuring Up? Trends and Achievement Based on the 1986 National Assessment	X						X		
The Underachieving Curriculum: Assessing U.S. School Mathematics from an International Perspective	X								
Saving the African American Child: A Report of the National Alliance of Black School Educators, Inc. Task Force on Black Academic and Cultural Excellence	X	X							
Who Will Do Science? Trends, and Their Causes, in Minority and Female Representation among Holders of Advanced Degrees in Science and Mathematics	X								

Categories

Title	Expectations	Equity	Learning Styles	Cooperative Learning	Practices & Programs	Funding Sources	Instructional Materials	Manipulatives	Computer Software
	Categories								
Handbook for Conducting Equity Activities in Mathematics Education	X	X	X						
Retention: Keeping Your Students in the '80s: How to Build a Successful Tutorial Program to Help You Maintain High Enrollments	X			X					
The Urban Challenge—Poverty and Race	X	X							
Twice as Less; Black English and the Performance of Black Students in Mathematics and Science	X		X		X				
Academic Preparation for College in Illinois: Admission Requirements for Public Colleges and Universities and Suggested Learning Outcomes for College-bound Students	X								
We Have a Choice: Students at Risk of Leaving Chicago Public Schools	X								
Overcoming Math Anxiety	X		X		X				
"School Desegregation Its Effects: An Introduction"	X	X							
Strengthening and Enlarging the Pool of Minority High School Graduates Prepared for Science and Engineering Career Options	X								
The Education Equality Project Academic Preparation in Mathematics: Teaching for Transition from High School to College	X								

Title	Expectations	Equity	Learning Styles	Cooperative Learning	Practices & Programs	Funding Sources	Instructional Materials	Manipulatives	Computer Software
Changing America: The New Face of Science and Engineering; Interim Report		X							
Black Children: Their Roots, Culture, and Learning Styles			X						
The Truly Disadvantaged: The Inner City, the Underclass, and Public Policy			X						
Educating Black Children: America's Challenge				X					
TEAM: Teaching for Excellent Achievement through Motivation				X					
Educational Equality Project: Academic Preparation for College: What Students Need to Know and Be Able to Do					X				
Mathematics and the Urban Child: A Review of Research. What do we know about teaching and learning in urban schools?					X				
Gruber's Super Diagnostic Tests for the SAT: Programmed to Critical Thinking Strategies to Maximize Your Scores					X		X		
Adolescent Programs that Work					X	X			
Effective Compensatory Education Sourcebook: Volume 1, a Review of Effective Educational Practices					X				
Effective Compensatory Education Sourcebook: Volume II, Project Profiles				X	X				

Categories

Title	Expectations	Equity	Learning Styles	Cooperative Learning	Practices & Programs	Funding Sources	Instructional Materials	Manipulatives	Computer Software
Institutional Projects Funded by OERI						X			
The Ideas of Algebra, K–12							X		
Problem Solving In Mathematics: Algebra							X		
Algebra Warm-Ups							X		
Algebridge: Assessment and Instruction Materials							X		
Algebra Word Problems							X		
Algebra in the Concrete							X		
Playing with Infinity: Mathematical Explorations and Excursions							X		
The 1979 Hammond Almanac							X		
The Nation's Report Card: Improving the Assessment of Student Achievement, Report of the Study Group							X		
The World Almanac and Book of Facts, 1977							X		
Hands-on Equations Math Lab Kit								X	
Algebra Tiles for the Overhead Projector								X	
Equations Game								X	
WFF'N Proof								X	
What's My Rule Game								X	
The Function Analyzer									X
Green Globs and Graphing Equations									X
Tobbs Learns Algebra									X
Keep Your Balance									X

Title	Expectations	Equity	Learning Styles	Cooperative Learning	Practices & Programs	Funding Sources	Instructional Materials	Manipulatives	Computer Software
MathCad									X
Trivia Math: Algebra								X	
Algebra Before and After: Book 1								X	
Algebra Tiles Activities: Book 1								X	
Algebra Tiles								X	
Rocky's Boots									X
Robot Odyssey									X
Super Plot									X
Middle Grades Mathematic Project								X	
Solid Sense in Mathematics 7–9								X	

ANNOTATED LIST OF RESOURCES

(Annotation appears with initial entry only)

Expectations

Expectations have been identified as critical to successful experiences in algebra, as well as in other areas of mathematics learning. Materials available in this area address the expectations of society, educators, parents, and the students themselves. The articles also discuss the interrelatedness of the aspects of various groups and their impacts on the mathematics education of minority students. The following resources are examples of documents that address this issue.

Beane, DeAnna B. *Mathematics and Science: Critical Filters for the Future*. Washington, D.C.: Mid-Atlantic Center for Race Equity, American University, 1985.

The booklet reviews the role of the principal in effecting change; furnishes data and background information about factors underlying the underrepresentation of blacks, Hispanics, and Native Americans in advanced mathematics and science courses; describes significant components of successful programs addressing underrepresentation; suggests materials that facilitate assessment and intervention planning; and presents resources that support prevention and intervention strategies at the local school level.

Berryman, S. E. *Who Will Do Science? Trends, and Their Causes, in Minority and Female Representation among Holders of Advanced Degrees in Science and Mathematics.* New York: n.d., Rockefeller Foundation.

Cheek, Helen Neely, Gilbert J. Cuevas, Judith E. Jacobs, Genevieve Knight, and B. Ross Taylor. *Handbook for Conducting Equity Activities in Mathematics Education.* Reston, Va.: National Council of Teachers of Mathematics, 1984.

The handbook includes suggestions for conducting mathematics equity surveys, designing and organizing equity conferences and other teacher in-service activities, developing networking strategies, and developing curriculum and instructional strategies that deal with equity issues in mathematics; lists resources of mathematics equity materials; and presents state-of-the-art papers on underrepresented groups in mathematics.

College Entrance Examination Board (CEEB). *The Education Equality Project Academic Preparation in Mathematics: Teaching for Transition from High School to College.* New York: CEEB, 1985.

This is the mathematics component of a series of books relating to *Academic Preparation for College: What Students Need to Know and Be Able to Do.* It is a dialogue to support teachers as they seek to implement the mathematics component of the Educational Equality Project.

Dossey, John A., Ina V. S. Mullis, Mary M. Lindquist, and Donald L. Chambers. *The Mathematics Report Card: Are We Measuring Up? Trends and Achievement Based on the 1986 National Assessment.* Report No.: 17-M-01. Princeton, N.J.: Educational Testing Service, 1988.

The report chronicles trends in proficiency across four mathematics assessments conducted in 1972–73, 1977–78, 1981–82, and 1985–86. Trends across four assessments involving 150 000 students since 1973 are summarized to offer a comprehensive view of achievement patterns for students aged nine, thirteen, and seventeen.

Fox, Linda, L. Brody, and Dianne Tobin. *Women and the Mathematical Mystique.* Baltimore: Johns Hopkins University Press, 1980.

The text is divided into four sections. The descriptive titles of the first three sections are (1) "Female Mathematicians," (2) "Sex Differences in Mathematics Achievement and Course Taking," and (3) "Facilitating Women's Achievements in Mathematics." The final section includes a summary.

Green, Robert L. "School Desegregation and Its Effects: An Introduction." *The Urban Review: Issues and Ideas in Public Education* [Journal of the National Urban Education Association] 13 (Summer 1981): pp. 51–56.

______. *The Urban Challenge—Poverty and Race.* Chicago: Follett, 1977.

The book presents the author's perceptions of urban problems and provides empirical data in support of his perceptions. It examines many dimensions of the urban dilemma and offers guidelines for future directions. The seventh chapter, "Education: Quality for All," includes a section titled "Tracked and Trapped," which sheds light on the importance of algebra for everyone.

Hilliard, Asa G., III, et al. *Saving the African American Child: A Report of the National Alliance of Black School Educators, Inc., Task Force on Black Academic and Cultural Excellence.* Washington, D.C. : National Alliance of Black School Educators, 1984.

The booklet sets forth a set of standards for quality education for the African-American child. The booklet defines the problem; specifies what kind of education is needed for equity, academic excellence, and cultural excellence; and defines the needs for organizations and resources to assure academic and cultural excellence for black students.

Howard, Jeffery, and Raymond Hammond, "Rumors of Inferiority: The Hidden Obstacles to Black Success." *New Republic* 193 (9 September 1985).

The author addresses the performance problem of blacks that results in large part in a tendency to avoid intellectual competition. He attributes the problem to a psychological phenomenon that arises when the larger society projects an image of black intellectual inferiority and when that image is internalized by black people. He discusses extensively intellectual development, the psychology of performance, and the commitment to development.

Illinois State Board of Higher Education and the Illinois Board of Higher Education. *Academic Preparation for College in Illinois: Admission Requirements for Public Colleges and Universities and Suggested Learning Outcomes for College-bound Students.* Illinois Springfield: State Board of Education, 1988.

This report of the Illinois joint task force on admissions requirements describes what high school graduates should know and be able to do if they plan to pursue a baccalaureate degree.

Johnson, D. "Strengthening and Enlarging the Pool of Minority High School Graduates Prepared for Science and Engineering Career Options." Prepared for the Black Caucus Legislative Weekend Symposium on Opportunities for Minorities in Science and Technology: Preparing for the Year 2000, September 1988.

The paper suggest strategies for mobilizing local communities and families in which the children live to develop solutions to the problem.

Kunjufu, Jawanza. *Countering the Conspiracy to Destroy Black Boys.* Chicago: African American Images, 1985.

The author discusses the societal factors that begin at childhood to handicap African American males, describes extensively the fourth-grade failure syndrome and male reasoning, and presents strategies for countering these debilitating forces.

———. *Countering the Conspiracy to Destroy Black Boys: Volume II.* Chicago: African American Images, 1986.

Volume 2 of this two-volume series looks at the impact that home and school can have on boys in more detail. The five chapters include "Developing Responsibility in Black Boys," "Female Teachers and Black Male Culture," "Some Reasons Why Black Boys Succeed or Fail," "A Relevant Curriculum for Black Boys," and "Simba" (Young lion).

———. *Developing Positive Self-Images and Discipline in Black Children.* Chicago: African American Images, 1984.

The author proposes a holistic approach to solving the problems related to positive images and self-discipline of black children. He then develops the book in six chapters with descriptive titles: (1) "The Politics of Educating Black Children," (2) "Developing Positive Self-Images and Self-Esteem in Black Children," (3) "A Relevant Curriculum," (4) "Developing Self-Discipline in Black Children," (5) "Parenting: Children Are the Reward of Life," and (6) "From Theory to Practice."

———. *Motivating and Preparing Black Youth to Work.* Chicago: African American Images, 1986.

This action-oriented book is developed in six chapters with descriptive titles: (1) "The Politics of Work," (2) "Values as the Foundation for Motivation," (3) "Motivation to Work," (4) "The Development of Talents into a Career," (5) "Jobs: The Present and Beyond," and (6) "How to Become Economically Self-Sufficient."

Kyle, C. L., J. Lane, J. A. Sween, and A. Triana. *We Have a Choice: Students at Risk of Leaving Chicago Public Schools.* A Report to the Chicago Board of Education and the Illinois Attorney General. Chicago: DePaul University, March 1986.

This report on students at risk of leaving Chicago Public Schools gives an empirical overview of key characteristics of students at risk of leaving high school.

Majer, Kenneth. *Retention: Keeping Your Students in the '80s: How to Build a Successful Tutorial Program to Help You Maintain High Enrollments*. San Diego: Consulting Group, 1980.

This handbook on developing and implementing a tutoring program was designed for self-instructional use or use in a four-day seminar that furnishes practical experiences necessary to develop and manage tutorial programs. The topics covered include writing instructional objectives; diagnosing students' learning problems; questioning techniques; problem-solving skills in mathematics and science; diagnosing study and organization problems and teaching study skills; avoiding communications barriers; psychological principles of teaching, tutoring, and learning; and evaluation of a tutoring program. Although the materials were designed for use in establishing tutoring programs for students whose backgrounds have not prepared them sufficiently well for university work, they have been used effectively for all mathematics courses at the high school level.

McKnight, Curtis C., F. Joe Crosswhite, John A. Dossey, Kenneth J. Travers, and Thomas J. Cooney. *The Underachieving Curriculum: Assessing U. S. School Mathematics from an International Perspective*. Champaign, Ill.: Stipes Publishing Co., 1987.

This book, which summarizes the major findings of the Second International Mathematics Study, provides empirical data on a great many aspects of school mathematics, including the content of the mathematics curriculum, what mathematics was taught by the teacher, how mathematics was taught, and students' achievement and attitudes.

Orr, Eleanor W. *Twice as Less: Black English and the Performance of Black Students in Mathematics and Science*. New York: W. W. Norton & Co., 1987

The author seeks to answer the question, Does black English stand between black students and success in mathematics and science? She reports on an experiment in a program in the District of Columbia in which she found that the performance of black students in mathematics and science is crippled not by lack of intelligence or diligence but by linguistic interference. She relates nonstandard English usage to misunderstanding of concepts.

Tobias, Sheila. *Overcoming Math Anxiety*. Boston: Houghton Mifflin Co., 1978.

The book is a discussion of how intimidation, myths, misunderstandings, and missed opportunities have affected a large proportion of the population. It is designed to convince women and men that their fear of mathematics is the result and not the cause of their negative experiences with mathematics and to encourage them to give themselves another chance.

Equity

Much research has been done that addresses the equity and opportunity-to-learn issue as it relates to mathematics education. The results of this inequity are documented. References in this section are examples that lend insight into this issue.

Cheek, Helen N., Gilbert J. Cuevas, Judith E. Jacobs, Genevieve Knight, and B. Ross Taylor. *Handbook for Conducting Equity Activities in Mathematics Education*. Reston, Va.: National Council of Teachers of Mathematics, 1984.

Green, Robert. *The Urban Challenge—Poverty and Race*. Chicago: Follett, 1977.

Hilliard, Asa G., III, et al. *Saving the African American Child: A Report of the National Alliance of Black School Educators, Inc. Task Force on Black Academic and Cultural Excellence*. Washington, D.C. : National Alliance of Black School Educators, 1984.

Howard, Jeffery, and Raymond Hammond. "Rumors of Inferiority: The Hidden Obstacles to Black Success." *New Republic* 193 (9 September 1985).

Reynolds, W., and J. Oaxaca. *Changing America: The New Face of Science and Engineering Interim Report.* Washington, D.C.: Task Force on Women, Minorities, and the Handicapped in Science and Technology, 1988.

The booklet summarizes a report to the president that includes six goals for the nation, graphical presentations of the problem related to each goal, and actions to be taken to achieve each goal.

Simmons, Cassandra, and Nelvia Brady. "School Desegregation and Its Effects: An Introduction." In Urban Review: Issues and Ideas in Public Education [Journal of the National Urban Education Association] 13 (Summer 1981): 51–56.

The article discusses student placement involving labeling, classifying, and differential assignment policies, noting that ethnic and racial minority students are disproportionately allocated to the lower-achieving academic groups. The importance of lowered teacher expectations for these groups is noted, and the educational consequences are described.

Learning Styles

Research in the last few years has focused on learning styles that are common to various cultures and their impact on the mathematics education of representatives of those cultures. Some of the research discusses the impact of ignoring the effects of culture on learning styles. They also highlight the danger of demanding instantaneous changes from culturally influenced learning styles to the learning styles of the dominant culture. The following references are examples of documents that address this issue.

Benson, Janet H. *Black Children: Their Roots, Culture, and Learning Styles.* Rev. ed. Baltimore: Johns Hopkins University Press, 1982.

The book uses research to highlight the significance of culture and cultural experiences on the development and educational advancement of Afro-American children in the United States. She defines the two noncongruent cultures that all black children must master and the third culture that black males must master.

Cheek, Helen N., Gilbert J. Cuevas, Judith E. Jacobs, Genevieve Knight, and B. Ross Taylor. *Handbook for Conducting Equity Activities in Mathematics Education.* Reston, Va.: National Council of Teachers of Mathematics, 1984.

Orr, Eleanor W. *Twice as Less: Black English and the Performance of Black Students in Mathematics and Science.* New York: W. W. Norton & Co., 1987.

Tobias, Sheila. *Overcoming Math Anxiety.* Boston: Houghton Mifflin Co., 1978.

Wilson, W. J. *The Truly Disadvantaged: The Inner City, the Underclass, and Public Policy.* Chicago: University of Chicago Press, 1987.

The two sections of the book, (Part 1) "The Ghetto Underclass, Poverty, and Social Dislocation," and (Part 2) "The Ghetto Underclass and Public Policy" discusses the social pathologies of the ghetto and offers a comprehensive explanation of the rise of inner-city problems. It then presents an agenda that moves beyond race-specific issues to confront fundamental problems of our society.

Cooperative Learning

Research has documented the effectiveness of cooperative learning on the mathematics achievement of students who are not always successful in classroom mathematics learning experiences. References in this section address both the need for, and effects of, cooperative learning and strategies for incorporating cooperative learning into instructional programs. The resources listed here also include the names of people who have successfully implemented cooperative learning in mathematics programs for underrepresented groups.

Jones, J. A. *TEAM: Teaching for Excellent Achievement through Motivation.* Silver Spring, Md.: Century Technologies, 1988.

This program is designed to help teachers motivate students in mathematics by tying the subject-matter curriculum more closely to individual students' interest, motivation, and learning styles and training teachers to use the more interesting subject matter and techniques to motivate students.

Majer, Kenneth. *Retention: Keeping Your Students in the '80s: How to Build a Successful Tutorial Program to Help You Maintain High Enrollments.* San Diego: Consulting Group, 1980.

Strickland, Dorothy, and Eric Cooper. *Educating Black Children: America's Challenge.* Washington, D.C.: Bureau of Education Research, School of Education, Howard University, 1987.

The book contains eleven essays on issues affecting the education of black children. The essay "Cooperative Learning and the Education of Black Students" presents guidelines for using cooperative learning strategies with black students.

Practices and Programs

Some programs are currently operational that effectively achieve success with mathematics education for students who have been identified as limited in their abilities to achieve desired proficiency in mathematics. Examples of such programs are included here.

Burke, F., and Ralph Lataille. *Adolescent Programs That Work.* Trenton: State of New Jersey, Department of Education, November 1978.

The booklet contains descriptive information on sixty-four innovative and successful educational programs developed throughout the nation, addressed to adolescent students in grades 6 to 12.

College Entrance Examination Board (CEEB). *Educational Equality Project: Academic Preparation for College: What Students Need to Know and Be Able to Do.* New York: CEEB, 1983.

The booklet presents a comprehensive description of the knowledge and skills needed by all college entrants in six basic academic competencies (reading, writing, speaking and listening, mathematics, reasoning, and studying) and six basic academic subjects (English, the arts, mathematics, science, social studies, and foreign language) to strengthen the academic quality of secondary education and to ensure equality of opportunity for postsecondary education for all students.

Griswold, Phillip A., Kathleen J. Cotton, and Joe B. Hansen. *Effective Compensatory Education Sourcebook, Volume I: A Review of Effective Educational Practices.* Washington, D.C. : U. S. Department of Education, 1985.

The booklet contains profiles of 118 Chapter 1 programs that were identified through extensive evaluation of thirteen attributes of effectiveness as being worthy of national recognition. Each profile contains a narrative summarizing three salient attributes of success and a chart listing essential descriptive information.

______. *Effective Compensatory Education Sourcebook, Volume II: Project Profiles.* Washington, D.C.: U. S. Department of Education, 1985.

Gruber, G. R. *Gruber's Super Diagnostic Tests for the SAT: Programmed to Critical Thinking Strategies to Maximize Your Scores.* New York: Barnes & Noble Books, 1988.

A set of diagnostic tests and instructions for preparing students for the SAT and other standardized tests. The book includes a strategy diagnostic-test section, a strategy-and-thinking-skills section, and a reference section.

Orr, Eleanor W. *Twice as Less: Black English and the Performance of Black Students in Mathematics and Science.* New York: W. W. Norton & Co., 1987.

Suydam, Marilyn. *Mathematics and the Urban Child: A Review of Research: What Do We Know about Teaching and Learning in Urban Schools?* St. Louis: CEMREL, 1978.

This monograph is one of a series of papers on the issues of instruction and learning in urban schools that were presented at a conference held 10–14 July 1978. The key problems to which the papers are addressed are assessment of learning outcomes and the analysis of the relationships between instructional and other inputs and learning outcomes.

Tobias, Sheila. *Overcoming Math Anxiety.* Boston: Houghton Mifflin Co., 1978.

Funding Sources

The list of funding resources for mathematics education for underrepresented groups is extensive. This section identifies potential funding sources that meet individual needs.

Burke, F., and Ralph Lataille. *Adolescent Programs That Work.* Trenton: State of New Jersey, Department of Education, November 1978.

U. S. Department of Education. *Institutional Projects Funded by OERI.* Washington, D.C.: U. S. Department of Education, Office of Educational Research and Development Information Services, April 1988.

Instructional Materials

Instructional materials available to support algebra instruction for all students are included in this list. Special efforts were made to identify materials that present concrete and representation models of algebraic concepts and that support a conceptual approach to the development of proficiency in mathematics concepts, skills, and problem solving. The list assumes that instruction must move from "not knowing to knowing" on many occasions.

Alexander, Lamar H., and Thomas James. *The Nation's Report Card: Improving the Assessment of Student Achievement: Report of the Study Group.* Washington, D.C.: National Academy of Education, 1987.

The book examines how we in the United States currently assess what our students across the country know and can do, and suggests ways of improving our progress. The booklet also includes a review of the report by a committee of the National Academy of Education.

Bacheller, M. *The 1979 Hammond Almanac.* Maplewood, N. J.: Hammond Almanac, 1978.

A compendium of information for developing mathematics applications with a "real world" flavor; includes information on such topics as sports, history, government, politics, health, and science.

College Entrance Examination Board (CEEB). *Algebridge: Assessment and Instruction Materials.* Providence, R.I.: Jason Publications, 1990.

This is a set of supplementary curriculum materials that aim at bridging the gap between arithmetic and algebra by helping teachers to assess and instruct algebraic thinking within the context of arithmetic. The program focuses on concepts from arithmetic that are needed in the study of algebra that are not ordinarily stressed in arithmetic; emphasizes developing concepts and abilities rather than memorizing algorithms; and seeks to diagnose and correct errors in thinking, not errors in computation. It is an extension of the College Board's Education Equality Project.

Coxford, Arthur F., ed. *The Ideas of Algebra, K–12.* 1988 Yearbook of the National Council of Teachers of Mathematics. Reston, Va.: The Council, 1988.

The book is organized in six parts. Part 1 discusses the forces impinging on algebra in the curriculum and suggests possible directions for change. Part 2 concentrates on concepts and teaching possibilities available prior to the formal introduction of algebra. Part 3 focuses on equations and expressions in algebra. Part 4 emphasizes suggestions for teaching and using word problems in algebra. Part 5 emphasizes the use of technology in the algebra classroom. Part 6 focuses on teacher-tested ideas for teaching algebra.

Delury, G., et al. *The World Almanac and Book of Facts, 1977.* Madison, Wis.: Madison Newspapers, 1977.

A compendium of information for developing mathematics applications with a "real world" flavor; includes information on such topics as sports, people, places, things, and vital statistics.

Dossey, John, Ina V. S. Mullis, Mary M. Lindquist, and Donald L Chambers. *The Mathematics Report Card: Are We Measuring Up? Trends and Achievement Based on the 1986 National Assessment.* Report no. 17-M-01. Princeton, N.J.: Educational Testing Service, 1988.

Gruber, G. R. *Gruber's Super Diagnostic Tests for the SAT: Programmed to Critical Thinking Strategies to Maximize Your Scores.* New York: Barnes & Noble, 1988.

Harnadek, Anita. *Algebra Word Problems.* Pacific Grove, Calif.: Midwest Publications, 1988.

This set of eleven booklets with a teacher's manual and detailed solutions includes problems to give needed practice for confidence, competence, and mastery in solving algebra word problems. The series titles are "How to Solve Algebra Word Problems," "Warm-Up," "Ages and Coins," "Mixtures," "Formulas, Rectangles, $D = rt$," "Percents and Work Rates," "Miscellaneous A-1 (easy)," " Miscellaneous B-1 (medium)," "Miscellaneous C-1 (hard)," "Diophantine Problems," and "Fund Time."

Lane County Mathematics Project. *Problem Solving in Mathematics: Algebra.* Palo Alto, Calif.: Dale Seymour Publications, 1983.

Book includes a guide to problem solving using the PSM program; lists of problem-solving skills and strategies; lessons that concentrate on building problem-solving skills; and the guided-discovery approach in solving problems in the areas of drill and practice, grade-level topics, and challenge activities.

Laycock, Mary, and Reuben Schadler. *Algebra in the Concrete.* Hayward, Calif.: Activities Resources Co., 1973.

The booklet models ways for students to use concrete materials in exploring sequences, building algebraic expressions, factoring trinomials, and solving linear equations and simultaneous linear equations.

McFadden, Scott. *Algebra Warm-Ups.* Palo Alto, Calif.: Dale Seymour Publications, 1983.

These reproducible sets of warm-up exercises, each of which can be done in about ten minutes, provide review or word problems, equations, and bonus questions; they are usable to introduce new concepts and problem strategies as well as review previous lessons.

Peter, Rosa. *Playing with Infinity: Mathematical Explorations and Excursions*. New York: Dover Publications, 1961.

The author discusses many important mathematical concepts without being technical or superficial. She writes with complete clarity on the whole range of topics from counting to mathematical logic. The book uses very little algebra and no mathematical formulas. The target audience is beginning mathematics students and people in the humanities and other such fields.

Manipulatives

From Educational Teaching Aids, 199 Carpenter Avenue, Wheeling, IL, 60090:

Borenson, Henry, Hands On Equations Math Lab Kit.

In a concrete approach to building understanding of the concepts involved in solving equations, students use game pieces physically to build a given algebraic equation and then solve it through manipulation of the pieces.

Algebra Tiles for the Overhead Projector

Seventy transparent pieces and instructions are included for building patterns on the overhead projector to model such algebraic concepts as the zero principle, addition and subtraction of polynomials, and factoring trinomials.

Equations Game

This game builds problem-solving and thinking skills as students use the rules for equations in competitions.

WFF'N Proof

This twenty-one-game kit teaches symbolic logic, the rules of inference, logical proof, and formal systems. The games graduate in difficulty. As students play the games they increase their skills in abstract reasoning and precise thinking.

What's My Rule Game

This book contains forty-five black-line- master activity sheets that encourage the use of problem-solving techniques to determine the function from sets of number pairs. Included are teaching suggestions and answers.

From Creative Publication, P.O. Box 375, Aurora, IL 60507:

Algebra Tiles

Sturdy one-inch-square tiles made of clear plastic with the numerals 0–9 printed on them in black. Each set contains ten tiles.

Balka, D. *Algebra Tiles Activities: Book 1*

With this set of hands-on activities students use numeral tiles 0–9 to complete algebraic expressions on self-correcting, reproducible worksheets. Lessons are designed to build both algebraic and problem-solving skills.

Gregory, J. *Algebra Before and After: Book 1*

These activities were designed for use in the informal introduction of new topics and the application of newly learned skills. "Before" activities build intuitive understanding before concepts are formally introduced. "After" activities employ the discovery of patterns and relationships to facilitate practice of newly learned algebraic skills.

Pederson, K. *Trivia Math: Algebra*

This series deals with rational and irrational numbers, quadratics, conic sections, and more.

From Cuisenaire Company of America, 12 Church Street, Box D, New Rochelle, NY 10802:

Laycock, Mary, and Margaret Smart. *Solid Sense in Mathematics 7–9*

This booklet presents activities for using Cuisenaire Metric Blocks to develop mathematics concepts. The topic titles are "Square Root," "Pythagoras," "Algebra," "Sequences," and "Graphing."

Lappan, Glenda, Mary Jane Winter, et al. *Middle Grades Mathematics Project*

A set of activity books and manipulatives useful for developing mathematics concepts. The titles of the booklets are "Mouse and Elephant: Measuring Growth," "Similarity and Equivalent Fractions," "Factors and Multiples," "Spatial Visualization," and " Probability." The kit includes enough manipulatives for a class of thirty students.

Computer Software

From Mathsoft, Inc., One Kendall Square, Cambridge, MA 02139:

MathCad

This calculating software program is designed for use with the ordinary PC. It displays equations as the user writes them; combines text, numbers, and graphics anywhere on the screen; uses traditional mathematics symbols; has a pop-up menu for on-line help; solves simultaneous equations; does full matrix operations; flags errors; converts units automatically; creates plots in variable sizes; does advanced calculations; displays log plots; allows the user to calculate on the PC; writes the equation using familiar mathematical notations; and combines formulas, text, and graphs.

From Midwest Visual Equipment Company, 6500 North Hamlin, Chicago, IL 60645:

Robot Odyssey

The development of skills in logic and scientific problem solving is supported as students develop strategies, analyze problems, and form and test hypotheses as they attempt to escape from Robotropolis—a futuristic underground city populated by robots.

Rocky's Boots

Logically thinking, creativity, and problem-solving skills are developed as students use these skills in building simple logic circuits and making Rocky the Raccoon "kick" the correct target. The program includes over forty games. Disks are available for Apple II family, IBM, and Commodore 64.

Super Plot

The program facilitates graphing polynomials and trigonometric, logarithmic, and exponential functions one to five at a time. The featured capabilities of the program include superimposing graphs to compare them; scrolling to investigate key areas of interest; zooming in or out on either axis or on both axes simultaneously; changing limits; and displaying functions in standard algebraic notations.

From Sunburst Communications, 39 Washington Avenue, Pleasantville, NY 10570-2898:

Dugdale, Sharon, and Kibbey. Green Globs and Graphing Equations

In the first activity, "Linear and Quadratic Graphs," students are given a graph and must write the equations. In "Green Globs," students enter equations to create graphs that will hit thirteen green globs scattered randomly on the grid. "Tracker" requires students to locate linear and quadratic graphs that are hidden in a coordinate plane and to determine the equations. "Equation Plotter" is a general utility program that can be used to graph any general functions entered by the students.

Function Analyzer, The

The program provides students with a tool to explore the relationships among symbolic expressions, graphs, and tables of values. It emphasizes functions as a central idea in algebra. Included is a built-in function generator.

Keep Your Balance

In this problem-solving program students use deductive reasoning to solve balance equations. The program teaches and reinforces the skills of estimating, generalizing, finding alternative solutions, and writing an equation. Three levels of difficulty are included.

Tobbs Learns Algebra

This computer program is designed to develop students' algebraic thinking, hypothesis-making, testing abilities, and problem-solving skills.